William Picard Stephens

Canoe and Boat Building

A complete manual for amateurs.

William Picard Stephens

Canoe and Boat Building
A complete manual for amateurs.

ISBN/EAN: 9783337413170

Printed in Europe, USA, Canada, Australia, Japan

Cover: Foto ©berggeist007 / pixelio.de

More available books at **www.hansebooks.com**

CANOE AND BOAT BUILDING.

A COMPLETE MANUAL FOR AMATEURS.

CONTAINING PLAIN AND COMPREHENSIVE DIRECTIONS
FOR THE CONSTRUCTION OF CANOES, ROW-
ING AND SAILING BOATS AND
HUNTING CRAFT.

BY

W. P. STEPHENS,

Canoeing Editor of Forest and Stream.

WITH NUMEROUS ILLUSTRATIONS AND FIFTY PLATES OF
WORKING DRAWINGS.

FOURTH EDITION.

REVISED AND EXTENDED TO DATE.

NEW YORK:
FOREST AND STREAM PUBLISHING CO.
1889.

Copyright,
FOREST AND STREAM PUBLISHING CO.
1889.

PREFACE.

THE character and object of this book are set forth on its title page. It is a manual designed for the practical assistance of those who wish to build their own canoes.

The number of boating men who find pleasure merely in sailing a boat is small compared with those who delight not only in handling, but as well in planning, building, improving or "tinkering" generally on their pet craft, and undoubtedly the latter derive the greater amount of pleasure from the sport. They not only feel a pride in the result of their work, but their pleasure goes on, independent of the seasons. No sooner do cold and ice interfere with sport afloat than the craft is hauled up, dismantled, and for the next half year becomes a source of unlimited pleasure to her owner—and a nuisance to his family and friends. We know one eminent canoeist who keeps a fine canoe in his cellar and feeds her on varnish and brass screws for fifty weeks of every year.

This class of boating men, to whom, by the way, most of the improvements in boats and sails are due, usually labor under great disadvantages. Their time for such work is limited; they have not the proper outfit of shop and tools, nor the practical knowledge and skill only acquired by the professional builder after years of careful and patient labor; and the latter as a class are unwilling to communicate freely what they have acquired with so much difficulty, and are seldom willing to assist the amateur, even with advice. His only other source of information is reading, and while there are books treating of the construction of large vessels, and others of the use of boats, there are none giving precisely the instructions needed by the beginner in boat building.

Having experienced most of the trials and mishaps that fall to the lot of the tyro, we offer in these pages such help as has proved of the greatest value to ourselves. To the professional builder, some of the instructions may seem elementary and unnecessary; but it must be remembered that we are not writing for him, who by long practice has acquired an accuracy of eye and dexterity, that enable him to shorten, or to dispense altogether with many of the operations described. We are writing for the amateur who, in default of this training, must make up for it by extra care and patience, even at the expense of time, and the methods given are those which have proved best adapted to his peculiar requirements.

Canoe building is treated in detail, as the processes involved are common to all boatbuilding, only requiring greater care and skill than ordinary work; and the principles, once mastered, may be applied to the construction of any of the simpler craft, such as rowboats and skiffs.

It has been impossible to give due credit to the originators for many of the devices and inventions described; but to all such we return thanks in behalf of the great army of amateur builders and sailors, in which we claim a place.

CONTENTS.

	Page.
Introduction,	7
Designing,	11
Model Making,	19
Laying Down,	21
Methods of Building,	27
Tools and Materials,	31
Building,	40
Wells,	57
Aprons,	62
Paddles,	67
Sails and Rigging,	70
Centerboards,	88
Rudders,	94
Tabernacles,	97
Tents and Beds,	100
Stoves and Lamps,	106
Canvas Canoes,	111
Boat Building,	115
Appendix,	123
Description of Plates,	137

INTRODUCTION.

THE word canoe has two distinct meanings, having been applied, for how long a time no one knows, to boats of long and narrow proportions, sharp at both ends and propelled by paddles held in the hand, *without a fixed fulcrum*, the crew facing forward. The members of this great family vary greatly in size and model, from the kayak of the Esquimau, to the long war canoes, 80 to 100 feet long, of the islands of the Pacific. Within the past twenty years the word has been applied in England and America in a more limited sense, to small craft used for racing, traveling and exploring, as well as the general purposes of a pleasure boat, the main essentials being those mentioned above, while sails and a deck are usually added, the double paddle being used exclusively. In Canada the term has for a long time been applied to a similar boat, used for hunting and fishing, without decks, and propelled by a single paddle. The following pages will refer only to the second meaning given, as the one of most importance to the amateur builder, and as the instructions given will apply equally to the simpler and less complicated Canadian open canoe.

The modern canoe which, although in use for some years previously, may be said to date from Mr. MacGregor's cruises and books, 1865, 6 and 7, was in its early years divided into two distinct classes, Rob Roy and Nautilus, to which a third, Ringleader, was afterward added, but the many changes and improvements have so multiplied the models, that such names as Nautilus, Pearl, Shadow, Jersey Blue, etc., convey no definite idea of the boat's model or dimension. There are now no less than nine widely different models named Nautilus, six named Pearl, the Jersey

Blue has changed entirely, and half a dozen builders each offer a different Shadow, while dozens of other models have sprung up, so that such a division is no longer possible.

Modern canoes may, however, be classed in a different manner, according to the relative proportions of their paddling and sailing qualities, thus:

Paddling Canoes—Propelled solely by paddle.

Sailable Paddling—Sail being used as auxiliary, as in the early Rob Roy.

Sailing and Paddling—Both qualities being about equal, as in most cruising canoes.

Paddleable Sailing—Fitted mainly for sailing, as the later English boats, the paddle being auxiliary.

Sailing—Larger boats for two or three, using oars as auxiliaries, as the Mersey canoes.

For racing purposes a different classification has been adopted here, which, with the English, is given in the Appendix.

The first point in building a canoe is to decide on the model and dimensions, and this each man must do for himself, considering carefully the purpose for which he will use his canoe, the water she will sail on, the load to be carried, and similar details. The designs given cover all the different classes of canoes, and from them one can be selected as a basis for modification and improvement, to suit the builder. The following general directions will aid the novice in deciding on the main features of his craft:

For small streams and rivers, where portages have to be made, and sailing is of but little importance, a canoe 14ft.x27in. is most commonly used. She should have a flat floor, little or no keel, ends well rounded, little sheer. For general cruising work under sail and paddle, a canoe 14ft.x 30in., with flat floor, good bearings, sternpost nearly upright, model full enough to carry crew and stores easily, a keel of 2 to 3in. or a centerboard. For large rivers, bays and open waters, a canoe 14ft.x33in. or 15ft.x31½in., fitted with a metal centerboard of greater or less weight. The tyro will be safe in following either of these types, according to his purpose, as they are the ones usually preferred by canoeists.

Any object floating in water will sink until it displaces a weight of water equal to its own weight, thus with a canoe, if the hull weighs 90lbs., fittings 13lbs., sails and spars 15lbs., crew 145lbs., and tent, stores, ets., 50lbs., the total weight being 313lbs., it will sink until it displaces 313 lbs. of water, or $\frac{313}{62.5}=5$ cubic feet, as one cubic foot of fresh water weighs 62.5lbs. If in salt water, the divisor would be 65, a foot of the latter being 2½lbs. heavier than fresh.

Now, if that portion of our canoe which is below her proposed waterline contains less than 5 cu. ft., through being cut away too much, the boat will sink deeper than was intended, diminishing the freeboard and increasing the draft. This fault is found in some of the smaller canoes with fine lines, as when loaded to their full capacity they sink so deep as to be hard to paddle, and unsafe in rough water. To guard against it, a rather full model is desirable for cruising, where stores, etc., must be carried, it being hardly necessary to calculate the displacement, as is done with larger boats.

If, in making a model, a block of wood be taken 14in. long, 2¼in. wide and ¼in. thick, or one-twelfth as large each way as the portion of a 14ft. canoe below water, it will contain 17.5 cu. in., and if our model, when cut from this block, contains but 5 cu. in., it will be $\frac{5}{17.5}$ or .28 of the original block. This fraction, .28, is called the coefficient of the displacement, and expresses the proportion between the bulk of the boat below water and a solid whose dimensions are the length on loadline, the beam on loadline, and the depth from loadline to the outside of the bottom next the keel. In yachts it varies from .25 to .50, the former being called "light displacement" and the latter "heavy displacement" boats.

The displacement can be obtained, if desired, by first weighing the entire block, and after cutting out the model weighing that also, the ratio of one to the other being the coefficient of displacement mentioned above.

In the first class of canoes referred to, it is important to have the draft as light as is possible, as they are used often in very shoal waters. If built with a flat floor they need not draw over 4½ or 5in., the keel adding about 1in. more. Canoes of the second class usually draw 6in. exclusive of keel, which varies from 1 to 3in., the latter being the extreme limit allowed by the Association rules. The larger canoes are mostly centerboard boats, and draw from 6 to 7in. with no outside keel. The draft should be decided on and the position of the waterline fixed in the design, and the canoe trimmed to it as nearly as possible at first, changes in the ballasting being afterward made if they seem necessary.

The freeboard is the distance from the water to the deck, and in most canoes it is less than it should be. The "least freeboard," or the distance from the water to the lowest point of the deck, may be 4, 5, and 6in. respectively for each of the classes.

The curve of the gunwale from the bow downward to the middle of the boat, and up again at the stern, is called the sheer. The height of the bow above the point where the freeboard is least, is usually 3in. in the first class of canoes, and 6 to 7in. in the latter two, the stern being about 2in. lower than the bow in each.

The rocker is the curve of the keel upward from a straight line, and should be about 2in. for a 14ft. boat.

The midship section is a section across the boat at its greatest beam, and on its shape the model of the boat largely depends. As a canoe must carry a comparatively heavy load on a light draft, and must sail with little ballast, a flat floor is desirable. The sides should be vertical or slightly flaring, the "tumble home" or rolling in of the upper streak detracting from stability, and being of no use.

The round of deck may be 3in. in a 27in. boat, and 3½in. in a 30 to 33in. boat, as a high crown adds greatly to the room below, frees the deck quickly of water, and no valid objection can be made to it.

DESIGNING.

TO go into the subject of designing at any length is outside of the scope of our present work, but a short description of the method of drawing and tools used, will enable the beginner to do all the work necessary for a small boat, and will also serve to introduce him to a most fascinating employment for his leisure hours, the importance of which to the intelligent and progressive yachtsman or boat sailor is now generally admitted.

The amateur will require a drawing-board, which for canoe work need be only a smooth piece of white pine three feet long, one foot six inches wide, and three-quarters of an inch thick; the lower and left hand edges being straight and at a right angle to each other; a T square about thirty-six inches long, one or two triangles of wood, or better of hard rubber, a pair of dividers with plain and pencil points, several ship curves of various patterns, scales and splines. These latter are long flexible strips of wood or rubber, and are used for drawing curves. They are usually held in place by lead weights at short intervals, but an easier and cheaper way is to confine them by small pins driven into the board. The best scales are those printed on strips of bristol board, eighteen inches long, costing twenty cents each. They may be had with any desired number of parts to the inch. The most convenient scale for a canoe drawing is two inches to the foot (one-sixth full size), or one and a half inches, in which case a common two-foot rule may be used, each division of one-eighth of an inch on which will represent one inch. For the sail plan the scale may be one-quarter of an inch to the foot.

A good paper for working drawings is the "roll detail paper" which is strong, buff in color and may be had of four or five feet in width and of any length. Some drawing pins are also needed to fasten the paper to the board, or if it is to remain there for some time, small copper tacks may be used, as the square and triangle will work over them more easily. A few pencils and an India rubber will complete the necessary outfit, a drawing pen being added if the drawings are to be inked in when completed, as they should be. If much work is proposed a few more curves may be added, a pair of small spacing dividers, bow pen and pencil.

Three views are always used in delineating a vessel, as shown in Plate I. These are called the sheer plan, half breadth plan, and body plan. The sheer plan is a vertical section, lengthwise of the boat, showing the curve of stem and stern, the rabbet lines, the sheer or deck line *a b c*, and the buttock lines, as curved lines; and the water lines, Nos. 1, 2, 3, 4, and the frame lines, 1 to 13, as straight lines.

The half breadth plan shows the width of one side of the boat at the deck and at each of the water lines, these lines being curved (as well as the diagonals Nos. 1 and 2), the frame and buttock lines being straight. The body plan shows the cross section at every frame line or square station (1 to 13); also, the line of the deck, *a b c*, as it appears from a point directly in front of the boat. The lines in the right-hand half (1 to **X**) are the sections of the forward body, and those to the left (**X** to 13) the after body. The water lines, buttock lines and diagonals are all straight in this plan.

The general type of canoe being decided on, we will make out a table of dimensions for reference in drafting, as follows:

DIMENSIONS OF CRUISING CANOE.

Length, over all	14 ft.
Beam, extreme	30 in.
Beam, at water line	29 in.
Draft of water	7 in.
Depth, water line to rabbet (distance fh)	5 in.
Depth of keel	2 in.
Freeboard, bow (distance $a\,e$)	11 in.
Freeboard, midships (distance $b\,f$)	5 in.
Freeboard, stern (distance $c\,g$)	9 in.
Sheer at bow	6 in.
Sheer at stern	4 in.
Round (or crown), of deck	3½in.
Thickness of plank and deck	¼in.
Keel, sided (thickness)	1 in.
Keel, moulded (depth)	2¼in.
Stem and stern, sided (thickness)	1 in.
Rake of stern post	2 in.

With the paper stretched as tightly as possible, and the board on a table of convenient height before us, the light coming from the upper left hand corner of the paper, we first draw a base line, A B, near the lower edge of the paper and in length equal to 14ft. on our scale, using the T square with its head held firmly against the left-hand edge of the board. Now starting at 0, the right-hand end of the base line, we lay off with the dividers 14 spaces of 1ft. each, numbering them from 1 to 14 as in the drawing, and, shifting the T square to the lower edge of the board, we draw vertical lines at each point of division, or 15 in all, prolonging them sufficiently to cross the sheer plan above.

Now at a distance from A B equal to half the extreme beam, in this case $\tfrac{30}{2}$ or 15in., we draw a horizontal line. Leaving a little space between the upper limit of the half breadth plan and the sheer plan, we draw the base line of the latter, C D, and parallel to it, and at any convenient distance apart, the water lines, Nos. 1, 2, etc., drawing in first the load water line at a distance fh, above C D, equal to 5in.

The other water lines, one above and two below the load water lines, are spaced 2in. apart as the most convenient division in this case. The middle buttock and bow lines, and any others that may be necessary, are now drawn in the half breadth and body plans, and the diagonals are also drawn in the latter.

To avoid confusion of the many lines necessary, it is well to draw these "construction lines," which are the frame work on which the drawing is constructed, in red; then when the drawing is completed, the water lines and diagonals in the half breadth plan are drawn in blue, the latter lines being broken, the former full. The remaining outlines are drawn in full black lines. The base line C D is supposed to pass through the lowest point of the hull of the boat, exclusive of keel, which point, in nearly all canoes, would be the bottom of the planking at midships, next the keel.

Having the paper laid off, we will begin with the sheer plan, laying off between stations 7 and 8 the least freeboard, bf or 5in., making a small circle to mark the place. Now at the bow we measure up $a\,e$ or 11in. from the water line to the deck line, at the same time measuring in the width that our stem is to be, outside of the rabbet, $1\frac{1}{4}$ in.; and similarly at the stern, measure up 9in. and in 1in. to the points a and c. Taking a long spline, we will lay it on the drawing so as to pass through these three spots, confining it by lead weights or by small pins on either side of it at each point. If it does not take a "fair" curve without any abrupt bends, other pins or weights must be added at various points until it is true and fair throughout, when the line may be drawn in with a pencil.

Next the outline of the bow, bottom of keel and stern may be drawn in with a spline or the curves, and also the rabbet line, showing the ending of the plank. The height of the crown of the deck at midships may also be laid off, and the middle line of the deck drawn. The center line of the midship section is E F, the manner of finding its position being given further on, and on each side of it at a distance equal to half the extreme beam, the perpendiculars $s\,s$ are drawn; then, using a small spline or a curve, the midship section is drawn, according to the taste of the designer, the line beginning at rabbet in the keel, and ending at the point b, which is, of course, as high above the water line as the corresponding point in the sheer plan. The midship section is completed by drawing in the other half, measuring with the dividers the breadths from E F on each water line, and transferring them to the opposite side, afterward drawing a curve through all the points thus found. The round of the deck may also be drawn in the body plan, joining the two extremes of the midship section.

Now proceeding to the half breadth plan we will first draw in the half breadth of the keel, stem and stern. In a keel canoe the breadths will be the same throughout, from $\frac{3}{4}$ to 1in., but in a centerboard boat the keel must be wider amidships, to allow room for the trunk. In this boat the width at the bow and stern is 1in., so we lay off $\frac{1}{2}$in. and draw a line parallel to A B, to represent the "half siding" of the keel, as it is called. The same distance is laid off on each side of E F in the body plan, being other views of the same line.

The keel being laid off, the half breadth at the deck is taken from the body plan and set off at X on the half breadth plan. The side line of the deck, of course, passes through this point, its ends meeting the side of the keel at the points a and c, the distances of these points from stations 0 and 14 respectively, being the widths of the stem piece and stern post outside of the planking. A spline is bent through the three points so as to give the desired fulness at bow and stern, and the "side line," or half breadth, on deck is drawn in.

The breadth on No. 2 water line is now laid off at X and the endings of the line determined by squaring down from those points in the sheer plan where No. 2 water line cuts the rabbet of bow and stern to the siding of the keel in the half breadth plan. To test it we will run in some of the intermediate sections in the body plan, beginning with No. 4.

Three points of the water line are now determined, and to obtain others we refer to Table I. in the Appendix and find first, that in most of the canoes there described the midship section is placed at about the middle of the loadline, which in our boat would be 2¼in. aft of Station 7, the length on loadline being 13ft. 4¼in., the fore body being 6ft. 8¼in. and after body 6ft. 8in. An inspection of the tables shows that the length of the "middle ordinate" ($k\,l$) in canoes of a medium type is about 37 per cent. of the beam at the water line. Taking 36½ per cent of 29in. we have 10½in. as the half breadth at the middle of the fore body

For purposes of comparison of the various canoes, a dividing buttock and body line is also used, being drawn in the body and half breadth plans, midway between the center and the extreme beam. The distances ($r\,s - t\,v$) of the intersections of this line with the load water line, afford a comparative measure of the degree of fullness of the boats, which for the bow ranges from 29 to 47 per cent. of the length of the fore body, and for the stern from 25 to 46 per cent. of the after body, the larger fraction, of course, indicating a finer boat. For the fore body we will take 36 per cent as an average of cruising boats, then 36 per cent. of 6ft. 8¼in.=2ft. 6in., which, laid off along the bow line from the fore side of the stem at water line, gives a point on the water line, and similarly, taking 40 per cent. (a rather large figure, but the boat in question has a very fine run) we have 40 per cent. of 6ft. 8in.=2ft. 9in., which is laid off from the after side of stern at $w\,l$. With these five points given a spline is readily set and the water line drawn in.

Turning now to the body plan (the right hand side of which represents the frames of the fore body, and the left those of the afterbody) the sheer or deck line, $a\,b\,c$, is drawn. The T square is laid across the board at the height of the stem; a in the sheer plan is squared across to the half-siding of the stem at a in the body plan, and similarly the heights at Stations 2, 4, 6, are squared in. Now the half breadth at Station 6 is taken from the half breadth plan with the dividers and set off to the right of E F at the proper height, then 4 and 2 are treated in the same manner, after which a curve is drawn from X through the spots to a, showing the deck line of the port side of the canoe, as it appears from a point directly in front, after which the line is drawn in the after body in the same manner. Of course this line gives the upper endings of all the frame lines, 1 to 13.

Only every other one of these is drawn in, the moulds thus being 2ft. apart, but by laying off the stations 1ft. apart, the bulkheads, masts, etc., are more easily located.

The lower ends of all frame lines will be on the side line of keel in body and half breadth plans, the heights being taken along the rabbet at each station in the sheer plan. Stations 4 and 10 are now completed, the breadths on the water line being transferred from the half breadth to the body plan, and curves drawn through the three points in each frame thus obtained.

Now the remaining water lines, Nos. 1, 3 and 4, may be drawn in the half breadth plan, their endings being found by squaring down from their intersections with the rabbet in the sheer plan and the breadths at 4, X and 10 being taken from the body plan. When all the water lines are fair, the frame lines at 4 and 10 being altered slightly, if necessary, to correspond, the remaining stations, 2, 6, 8 and 12, may be completed.

The design is now ready for the final fairing, for which the "diagonals" No. 1 and No. 2 are drawn in the body plan. These lines should be so drawn as to intersect all the frame lines at as near a right angle as possible. The distances along the diagonal from the point i to the intersection of each frame line, are taken off in turn, and laid off on their corresponding stations in the half breadth plan, and a line is drawn through the points. If the line is unfair it must be altered, the corresponding points in the water and frame lines being changed at the same time, until all coincide, the breadths and heights of every intersection being the same in all three plans, when it may be assumed that the drawing is fair.

The endings of the diagonals are found by squaring across from the points in the body plan where they cross the siding of stem and stern to the rabbet line on stem and stern in the sheer plan, and then squaring down these points to the siding in the half breadth plan. The diagonals may be laid off in two ways, either an "expanded," as already described, or a "level" diagonal, in which the distances from E F in the body plan to each intersection are measured horizontally as $q\,d$.

As an additional test of fairness other "buttock" lines may be run in. These are drawn in the body and half breadth plans, parallel to the center lines, and are transferred to the sheer plan by taking the height of each intersection in the body plan and setting it off on the corresponding station, the curve being drawn through the "spots" afterward. The endings of the buttock lines are found by squaring up from the points in the half breadth plan where they cross the deck line, to the deck line in the sheer plan.

The process of "fairing" may be considered as completed when all the curved lines are true and fair, and the heights and breadths of every intersection are the same in each of the three plans.

This completes the "construction drawing," from which the calculations, if any, are made. Plate II. represents the completed "working drawing" of the same canoe, showing dimensions of keel, ribs, etc., and the position of all fittings. This may be a separate drawing, or the details may be added to the "construction drawing," after which all lines are inked in, as before directed.

MODEL MAKING.

IF the method of designing described be followed there will be no necessity for a model, but unless the amateur has had some practice in drafting it will be easier for him to first make a model, shaping it by eye, and then to take the lines from it.

In this case the design will be started on paper, as previously described, the sheer plan completed and the deck line drawn in the half breadth plan. To make the model, a block of soft dry white pine is required of a size to correspond with the scale of the drawing. The portion below the water line will be made of several thicknesses of pine and walnut or mahogany placed alternately. Each piece will

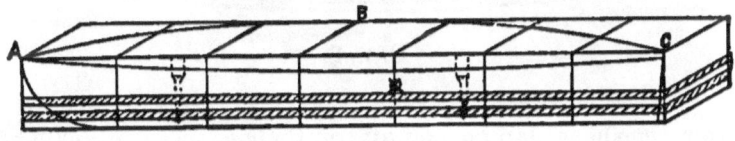

Fig. 1.

be of the same thickness as the distance between the water lines. A sufficient number of these pieces are taken to make up the required depth, and are fastened together with a few screws so placed near the back as not to interfere with the shaping of the model, and on top of all the pine piece is screwed, making a block like Fig. 1.

This block is now laid on the drawing, each of the divisions marked on it and numbered and the lines drawn on each of the four sides. On the back of the block the sheer plan is drawn, omitting the keel, if any, which will be added afterward. The half breadths on deck are now taken with the dividers from the half breadth plan and transferred to the top of the block, the curve run in with a spline pinned to the spots, and the block is cut to the line A B C.

The lines on the bottom of the block are now squared up across the new face, Fig. 2, the heights of the deck line taken

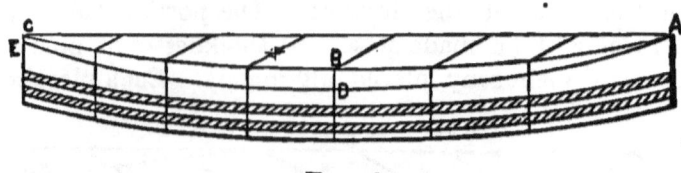

Fig. 2.

from the sheer plan and set off on their proper stations, and a line A D E drawn through the spots. The wood is now cut away to this line on the curved side and the line representing the middle of the deck on the back, leaving of course the same crown to the block as it is intended that the deck shall have. Next the back of the block is cut to the lines of the bow, stern and bottom.

The block is now screwed fast to a piece of board, which latter is nailed or screwed to the work-bench, so as to hold it firmly, and cut to the shape desired by the builder, the midship section being first roughed out, then the ends being cut away, and all finally finished off with sandpaper. To complete the model it is mounted on a board, the keel added, being glued to the board, it is varnished, and screw eyes put in to hang it up by. The model itself should be screwed to the board so as to be easily removed.

To take the lines from the block model the pieces are unscrewed and each laid in turn on the half breadth plan and the edge traced thereon, giving the water lines, from which the body plan may be drawn in. If it is desired to make a model from a drawing already completed, the operation is reversed, the pieces or "lifts" are planed to the proper thickness, the stations laid off across each, and the half breadths set off, the curve of the water line drawn, and the piece trimmed away to the line. After a piece is prepared for each water line, all are screwed together and the edges rounded off, forming a fair surface.

It is sometimes necessary to take the lines from a *solid* model, to do which the sheer plan is traced on paper, the half breadths on deck, and the heights to deck line are taken off by the dividers, and the lines drawn on the paper, then the frame lines are obtained by bending a thin strip of lead around the model and tracing the outline of it on the paper in the proper position, shown by the deck and sheer lines. The drawing being completed, the next operation is laying down.

LAYING DOWN.

This is the enlarging of the drawing to the full size of the boat, and is necessary in all boat and ship building. For large vessels, the drawing is made on the floor of the "mould loft," either in one continuous length, or in the case of very long vessels, in two portions overlapping each other. For small work a wide board will answer, 16ft. long and 2ft. wide, or less, according to the size of the boat.

On this common roll drawing paper is laid and tacked, and it is divided off, as was described for the detail drawing, except that the half breadth plan will now overlap the sheer plan to save room. Referring now to the detail drawing, a "table of offsets" is made. A sheet of paper is ruled in vertical columns, one for each square station, and also in horizontal lines as follows. In this table is set down the heights above base line of the stem, rabbet and sheer, and the half breadth at the deck, on each waterline, and on the diagonals:

TABLE OF OFFSETS, CANOE "JERSEY BLUE."

Stations	0	1	2	3	4	5	6	7	8	9	10	11	12	13	14
Height from Baseline to Gunwale	16	14½	13	11¾	11	10½	10	10	10	10⅛	10½	11¼	12	13	14
Half Breadths. Deck	7-16	4½	8	11	13	14⅛	14¾	15	14¾	14½	12¾	10⅞	8	4⅜	7-16
8-inch	...	2½	6¼	9½	12¼	14⅛	14¾	15	14¾	14½	12½	9⅝	6⅜	2¾	7-16
6-inch	...	1½	4⅝	8⅜	11½	13¼	14¼	14¾	14½	13⅝	12	8¾	5	2	7-16
4-inch	...	¾	3¼	6¾	10	12	13¼	13¾	13⅝	12⅝	10½	7	5⅝	1⅛	7-16
2-inch	...	...	1¾	4⅝	7¾	10⅛	11½	12	11¾	10⅜	8	4⅝	1¾	½	7-16

From this table the lines are laid down full size on the paper, each distance being measured off on its proper frame or water line, and a long, thin batten of pine run through the spots thus found. As we are now working from a smaller scale to a larger, all errors are increased in the same ratio, and though the small drawing may have been accurate there will be some errors in the large one, and to correct these the same process of "fairing" is necessary, as was before described; running in the water lines, frame lines, and diagonals with the battens until all the curved lines are fair and regular, and the breadths and heights of every point are the same in all three plans. When the drawing is faired the remaining details, such as masts, bulkheads, floor, etc, are drawn in their proper places.

The lines of the drawing now show the outside surface of the plank, but the moulds over which the boat is built must, of course, correspond with the inner surface of planking. In large work the model is often made to the outside of the frames only, then the breadths, when taken off, show the actual size of the frame. If the working drawings are made to include the plank, the thickness of the latter is deducted at some stage of the drafting prior to laying down. In our canoe, for convenience, the drawings will all include the plank, so in making the moulds its thickness, $\frac{1}{4}$in., must be deducted.

To copy the frame lines, a piece of thin board or cardboard A B C D, Plate XIX., is slipped under the paper of the large drawing, adjusted under the line to be copied, and held in place by a couple of tacks. Setting the points of the compasses $\frac{1}{4}$in. apart, a row of spots is pricked through the paper into the board, $\frac{1}{4}$in. inside the frame line, shown by the small circles in Fig. 3. At the same time points on the center line, E F, load water line and the diagonals D1 and D2 are also marked. The board is then removed, a batten run through the spots, and the wood trimmed away to the line. If the drawing is made on a board or floor the lines may be taken off, as in Fig. 4.

A batten about $\frac{3}{4} \times \frac{1}{4}$in. is bent along the line on the floor and held down by flat-headed nails. A piece of board is laid on top of the batten and a mark scratched on its under side with the piece of bent wire shown at A. In this case, after cutting to the mark another line must be gauged $\frac{1}{4}$in. inside the edge, and a second cut made to it, after which it is laid on the drawing and the center line, water line and diagonals laid off on it.

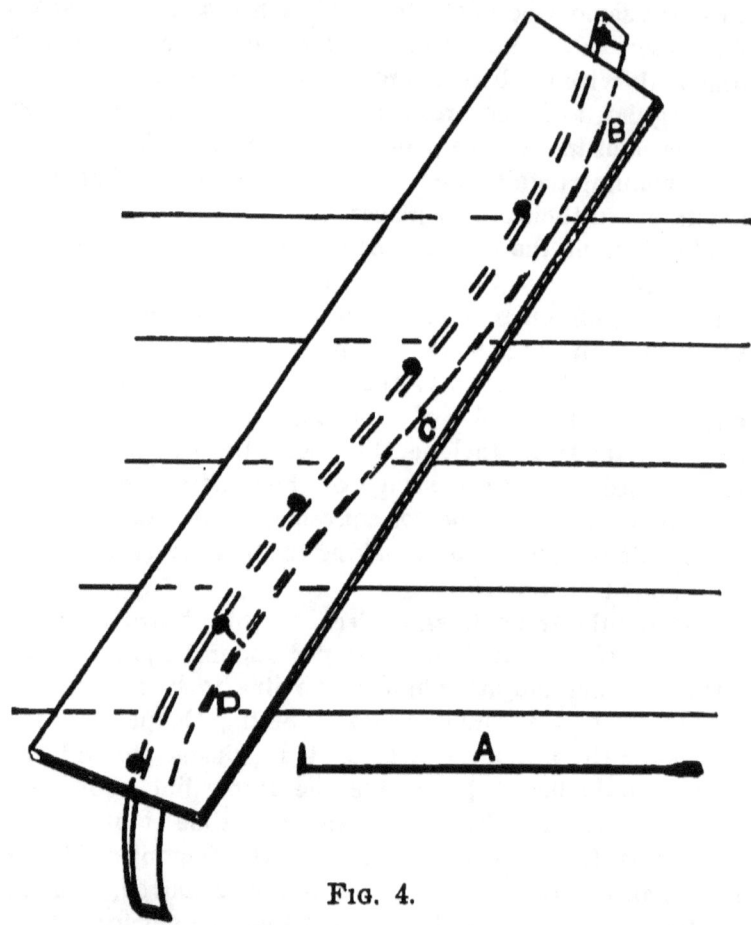

FIG. 4.

To make the complete mould, a piece of 1in. pine is planed up on one edge, H I, Fig. 5. a center line E F, is drawn at right angles to it, and also the load water line, then the pattern is laid on this board, adjusted to the center and water lines, and one-half marked off; then the pattern is turned over, adjusted on the other side of E F, and that side also marked off, the diagonals being marked at the same time.

AMATEUR CANOE BUILDING. 25

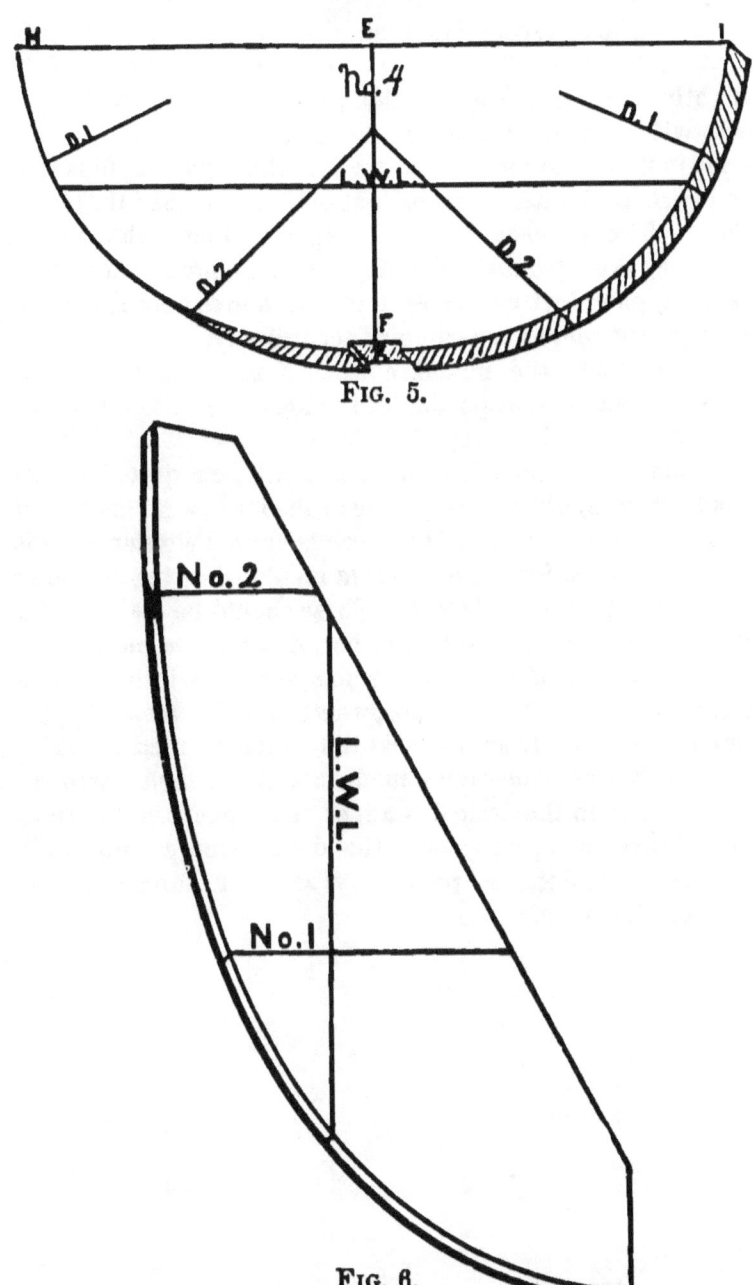

Fig. 5.

Fig. 6.

As the boat tapers from midships to the ends, it will be evident that the after side of the forward moulds will be slightly larger than the fore sides, and the reverse will be the case with the after moulds, No. X having both sides the same in most canoes. To allow for this bevel, moulds 2, 4, and perhaps 6 must be sawed out ¼in. larger than the marks show. The bevels at the deck height and on each diagonal are now taken from the drawing with a common carpenter's bevel, applied in turn to each of the above points, and the edges of the mould are trimmed accordingly.

To complete the mould, a notch K must be cut at the bottom to admit that portion of the keel or keelson inside of the rabbet, as will be explained later.

Besides the moulds described there will be required a stem mould (Fig. 6) giving the outline of the stem, a rabbet mould made to the rabbet line (if the stern is curved similar moulds will be required for it) and a beam mould, showing the curve and depth of the deck beams. These should be made of ½in. pine. They are taken off by either of the above methods.

The tendency of light boats is to spread in width in building, to avoid which in a canoe, where a small excess of beam may bar the boat from her class in racing, the model and all the drawings are sometimes made about one inch narrower amidships than the desired beam of the canoe, and the sides are allowed to spread when the deck beams are put in, if they have not done so previously, as often happens unless great care is taken.

METHODS OF BUILDING.

WHILE but few of the many different methods of building are adapted to the purpose of the amateur, a description of the principal ones will enable him to understand the entire subject more clearly. Of these, two are by far the most common, the carvel, and the lapstreak, also called clinker or clincher. In the first, usually employed for ships' boats, yawls, Whitehall and other boats, where lightness is not of first importance, the planks (six to eight on each side) are laid edge to edge, not overlapping, and nailed to the ribs or timbers that make the frame, the latter being spaced from nine to fifteen inches apart. To prevent leakage, a small thread of raw cotton, lamp wick, or in large boats, oakum, is driven into the seams with a mallet and caulking chisel, and the seams afterward filled with putty, marine glue, or if oakum is used, with pitch. To stand the strain of caulking and to hold the cotton, the planks must be at least $\frac{1}{2}$in. thick, which would be too heavy for a canoe.

In a lapstreak boat the planks lap over each other a distance of one-half to one inch, the edges being held together by rivets, some of these also passing through the ribs. In all cases the upper board laps on the outside of the one below it. Three objections are made to this mode of building—liability to leakage, difficulty of cleaning inside, and the obstruction that the laps offer to the water.

As to the first, it is almost entirely dependent on the skill and care used in the construction, and although a lapstreak boat may sometimes leak when first put in the water after drying out for a long time, it will very soon be perfectly tight. While the second point is an objection, it is by no means a serious one, and with a little care the boat may be kept perfectly clean, if not, a stream of water from a hose will wash out all dirt. The third point is the one most emphasized by the opponents of the lapstreak, but they overlook the fact that the laps, or lands, as they are usually called in England, are very nearly parallel, not with the water lines, but with the course of the water, which is largely down and under the boat. At the ends the lands are diminished to nothing, if the boat is properly built, and that they detract nothing from the speed is well proved by the fact that a very large majority of all canoe races have been won by lapstreak boats.

As to their advantages, they are light, easily repaired when damaged and they will stand harder and rougher usage than any other boats of their weight without injury. The lands on the bottom protect it greatly when ashore, and if anything they add slightly to the initial stability.

The oyster skiffs of Staten Island Sound and Princess Bay, boats from 18 to 25ft. long, lapstreak, of $\frac{3}{8}$in. plank, are considered by the fishermen to be stiffer and to rise more quickly than smooth-built boats of the same model. As after some experience with different modes of building, we have settled on the lapstreak as the best for canoes, and the easiest for amateurs, we shall later on describe it in detail.

In order to obtain a smooth skin, canoes are sometimes carvel built, as before described, but of $\frac{1}{4}$in. stuff, and as this cannot be caulked, a strip of wood about $\frac{1}{4}$in. thick and 1in. wide, is placed on the inside of each seam between the timbers, the edges of the planks being nailed to it. This is called the "rib and batten" plan, and is largely used in Canada.

Another and similar plan, the ribbon, or more properly ribband carvel (not "rib and carvel") is used in Massachusetts and Connecticut for whaleboats, and in England for canoes. In these boats the ribbands are of oak or ash, $1\frac{1}{4}$x $\frac{1}{4}$in., slightly rounded on the back and as long as the boat. They are screwed to the moulds, when the latter are in position, just where the seams of the planks come, and as each plank is laid on, its edges are nailed to the ribbands for their entire length. When the ribs are put in they must be "jogged" or notched over the ribbands. In both of these methods the boat is improved if a strip of varnished or painted muslin is laid along the seam, under the ribband, but this is often difficult to do. In a similar way the boats of the yacht Triton are smooth built, with a strip of brass inside each seam instead of a ribband of wood. While having a very fine surface these boats are usually not as tight as the lapstreak, and are more easily damaged.

In another method sometimes employed for canoes, the skin is double, the boat being first planked with $\frac{1}{4}$in. boards and then with a second layer, crossing the inner one. The first layer sometimes is laid diagonally, sloping aft from bow to stern with the second layer sloping the other way, so as to cross it nearly at right angles; a method used in U. S. Navy launches and lifeboats.

Sometimes the inner skin runs across the boat, and the outer fore and aft, as in the well-known "Herald" canoes, and sometimes both run fore and aft, the seams of one skin coming in the centers of the planks of the other, rivets being placed along all the edges, a method of building followed also in some of our largest cutter yachts.

With either of these methods a thickness of muslin is laid in paint between the two skins, and both are well nailed together. While making a very strong boat, it is often heavy, and when water once penetrates between the skins, as it will in time (with the thin plank used in boat building), the leaks cannot be stopped, and the wood will soon rot. Another serious objection to it is the great difficulty of making repairs.

Boats and canoes are sometimes built of tin, copper or galvanized iron, soldered and riveted together, a method usually confined to ships' boats and lifeboats. Two tin canoes were present at the first meet in 1880, and seemed strong, light and serviceable, though of poor shape. No doubt an excellent canoe could be built of sheet copper, that would not leak, and would be indestructible; but the cost and weight would be considerable.

In order to obtain a smooth skin with the advantage of the

lapstreak, the planks are sometimes rabbeted on their adjoining edges, half the thickness being taken from each plank, leaving smooth surfaces, inside and out, but thicker plank must be used than in the lapstreak, and the working is more difficult. In another mode the planking is in narrow strips, perhaps 1x⅜in. One of these is laid in place and nailed through from edge to edge, into the keel, then another is laid alongside of it and nailed to it, and so in succession until the boat is completed. A few frames are needed to stiffen the boat near the masts.

In the boats made by the Ontario Boat Company these strips are tongued and grooved, then steamed and forced together, the strips in some boats running fore and aft, and in others running around the boat, from gunwale to gunwale. In shell boats, where a very fine surface is of much greater importance than in canoes, the skin is made of Spanish cedar, about ¼in. thick, laid in four or six pieces, joining on the keel, and once or twice in the length of the boat, making one longitudinal seam and one or two transverse ones; but this method is not strong enough for canoes.

Paper has been used for the past thirteen years as a material for canoes, but although the boats are strong, tight, and but little heavier than the lapstreak, they have not become popular, and are but little used. The process of construction is patented, and requires both tools and experience beyond the reach of the amateur. Canoes have been built during the last five years on a similar system, using thin veneer in three thicknesses instead of paper, but, besides their great weight, no glue or cement can be depended on when long immersed in water; they are open to the same objections as all double-skinned boats, it is only a matter of time before leakage begins, after which they are practically ruined.

One of the oldest modes of boat building was to make a frame of wickerwork or similar material, covering it with leather, a method still followed, except that canvas is substituted for the leather. This mode of building is perhaps the easiest of all for the amateur, and we shall devote a chapter specially to it further on.

TOOLS AND MATERIALS.

In small boats, where lightness and strength are of first importance, it is necessary that the material should be very carefully selected, both as to quality and as to the fitness of each kind for the required purpose. Beginning with the keel, the best wood is white oak, with a clear, straight grain. In planing it will be found that the grain of the wood in one direction splinters and roughs up, while in the other it lies smooth and the keel should be so placed in the boat that the splinters or rough ends point *aft*, otherwise it will be torn in dragging over rocks and rough ground. In looking at the end of the wood, a series of concentric layers will be noticed. The piece should, if possible, be placed in such a position that the nails in it will pass *through* the layers, and not between two of them, for instance, in a keel the nails will be mostly vertical, so the layers of the wood should lie horizontally, and the same is true of the ribs, the nails through them being at right angles to the length of the boat, and the layers in each rib running fore and aft, thus avoiding any liability to split. Next to oak, either ash or yellow pine will make a good keel, but hickory should never be used in a boat, as it decays rapidly.

For the stem and stern, which are usually curved, the best material by far is hackmatack, or as it is sometimes called, tamarack, which may be had in knees of almost any curvature, from three to ten inches thick, or larger. For canoes a three-inch knee is the best, as if of full thickness it may be sawn into three slabs, each of which will make a stem and stern. Oak knees are also used, and are very good, but heavier. If knees cannot be had, the stem and stern may be cut out of straight plank.

For the sides of a centerboard trunk, clear, dry white pine is good, but mahogany is better, though much more costly. The timbers or ribs are usually of oak, though elm is excellent for this purpose. The wood must be clear and of the best quality in order to bend easily. The best oak for this purpose is found in the shape of stave timber used by coopers for the staves of barrels. These pieces are from three to five feet long, and about two by five inches square, one being sufficient for an ordinary lapstreak boat.

For planking, the very best material is white cedar, varieties of which are found along the entire length of the Atlantic seaboard. It is usually sold in boards $\frac{3}{4}$, 1 and $1\frac{1}{4}$in. thick for boat work, and from 12 to 20 feet long. For small boats it should be clear from sap and knots, but for larger work that is painted, the latter, if hard and sound, do not matter much, in fact, the knotty cedar is considered tougher and stronger than the clear.

Where cedar cannot be had, white pine can be used to advantage; in fact, the amateur will often find it much easier to buy pine of $\frac{1}{4}$in. already planed than to work up the thick cedar himself, while pine is not so apt to change its shape in working, a source of much trouble with cedar. Where neither of these can be had spruce may be used, but it is inferior. Mahogany and Spanish cedar make excellent planking, but they are no better than white cedar and cost much more. Most of the English books on canoeing recommend oak for planking, but it is never used here, being too heavy.

For the bulkheads, floor boards and inside work white pine is the best; for decks, rudder and upper streak of planking, mahogany, and for deck beams and carlings, spruce. The gunwale may be of spruce or pine, or, if outside, as will be shown, of mahogany, oak or yellow pine, the coamings of the cockpit being of oak. Paddles and spars are made either of white pine or spruce, the latter being stiffer and stronger, but a little heavier.

The other necessary materials—nails, screws, metal work, etc.—will be mentioned in detail as are required.

AMATEUR CANOE BUILDING.

The excellence of amateur work depends not, as many imagine, on the number of tools at hand, but on the care and perseverance devoted to it. The best work may be done with very few tools; but, on the other hand, it can be done much more quickly with a larger number.

If the amateur desires to build but one boat, at as small an outlay for tools as possible, the following will be sufficient:

Panel saw, 16in., 8 teeth to the inch.....	$1 00
Rip saw, 28in., 5 teeth to the inch........	2 00
Compass saw, 12in.......	40
Jack plane, double iron....	1 00
Smoothing plane, double iron........	85
Thumb plane...........	25
Claw hammer....	75
Riveting hammer........	40
Cutting pliers, Stubbs's or Hall's.......85c. to	1 25
Small screwdriver.......	50
Three gimlets, 1-13, ⅛, ¼in ...	50
Three brad awls.......	25
Six-inch try-square...........	35
Spokeshave........	50
Marking gauge.........	10
Chisels, ⅛, ½, 1in.......	75
Two-foot rule	25
Gouges, ⅛–1in.. inside bevel....	50
Oilstone...	1 00
Compasses, 5in.......	40
Four iron clamps, 4in........	2 00
Chalk line and scratch awl.........	25
	$15 25

The above are about the prices of the best quality tools, cheap ones not being worth buying, and with them any kind of small boat can be built, but the addition of the following tools will save some time and trouble:

Eight-inch ratchet brace	$1 85
Center and German bits, various sizes	1 50
Countersink....	25
Rabbet plane........	60
Bead plane, one-quarter inch........	50
Draw knife, nine-inch wide blade..........	1 50
Screwdriver, ten-inch	05
Twenty-six-inch hand saw ⎰ Instead of sixteen- ⎱	1 75
Eight-inch back saw ⎱ inch panel saw........ ⎰	1 10

These will be all that are needed, except a few files, and two or three drills to fit the brace, for the brasswork, such as the stemband, but there are some others that are very useful, though by no means indispensable, as follows:

Two-foot steel square.
Bench axe.
Expansion bit, seven-eighths to three-inch.
Level.
Convex spokeshave, for oars and paddles.
Mortise gauge.
Adze, for larger boats.
Small hand-drill stock with drills.
Two or three round sole planes for spars.

Besides these tools there will be needed a block of iron called a "set," or riveting iron, used to hold against the head of a nail in riveting; a "burr starter," which is a piece of iron or brass rod $\frac{3}{8}$in. in diameter and 3in. long, with a small hole in one end, used to drive the burrs on to the nails, and some wooden clamps, shown in Fig. 7. The solid ones are sawed out of oak, 6 to 8in. long and 1in. thick, strengthened by a rivet through them. The others are of the same size, but in two pieces, joined by a bolt or rivet. In use a wedge is driven in the back, closing the other ends of the jaws.

A work bench of some kind must be had, the simplest form being a plank 2in. thick, 10in. wide, and, if possible, several feet longer than the intended boat, so as to allow room for a vise on one end, as well as space to plane up long boards. This plank should be securely fastened along a wall, 2ft. 8in. above the floor and with its outer edge 20in. from the wall, the space at the back being filled in with 1in. boards, making a bench 20in. wide, the top being level and smooth, as the material to be planed on it will be very thin. A vise of some kind must be placed near the left hand end, an iron one being the best, but the common wooden one will answer, and is much cheaper.

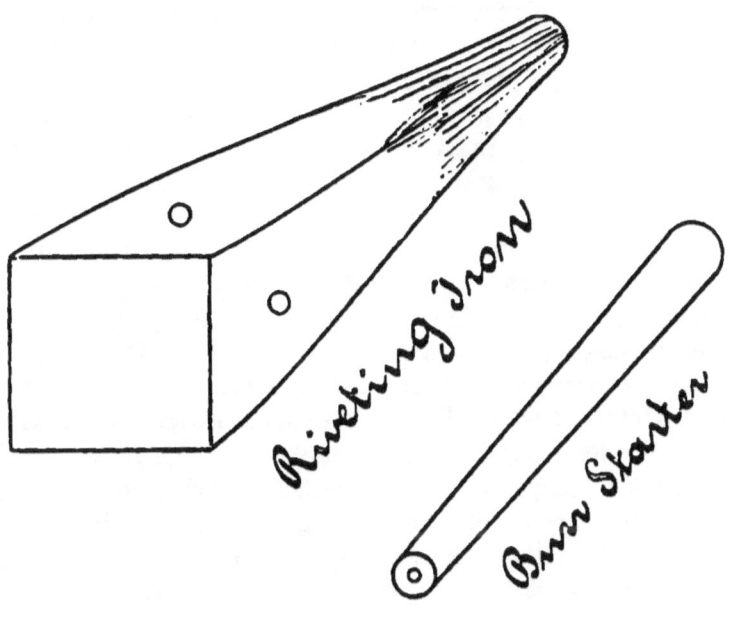
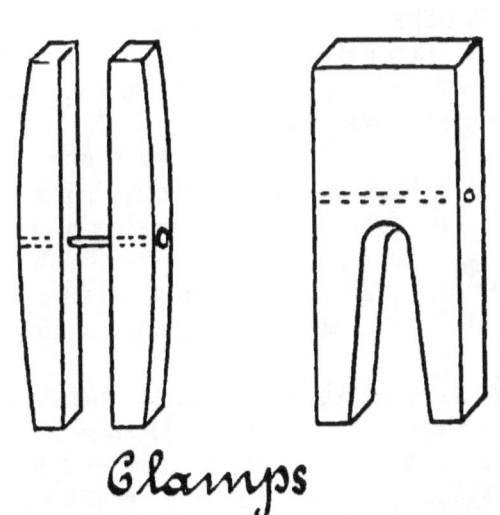

Fig. 7.

Fig. 8 shows a permanent bench fastened to the wall. The top is 3in. thick, of oak, and should be 24in. wide, and at least 10ft. long, a piece of 2in. plank being fastened at the right hand end by way of an extension for planing long stuff. A series of ¾in. holes about 3in. apart are bored in each leg, a peg being inserted in one of them to support long boards, in planing the edges. A bench hook (a) is placed near the vise; the bracket (c) is cut out of 2in. stuff and is bolted to the bench, being used to support spars, paddles and similar pieces, one end being held in the vise, and the other resting on the bracket.

Drawers are provided under the bench for tools, nails, screws, etc. At the back of the bench an upright board 12in. wide, carries a rack for the chisels, gouges, gimlets and small tools, above it, on the wall, the saws, draw-knife, spokeshave, brace, etc., are hung, a rack for the small planes, and another for sandpaper is fastened, also small boxes for such nails and screws as are most frequently required.

Two saw horses or benches are also necessary, the tops being 3in. thick, 6in. wide and 3ft. long, and the legs 2ft. long. Two pins of hard wood 1in. in diameter are driven tightly into holes about 1¼in. apart in one of the benches. When not in use they are driven down flush with the top, but in slitting long boards, they are driven up and the board wedged between them.

Another useful piece of furniture is a stool about 1ft. x 18in. on top and 18in. high, one-half of the top being a seat and the other half, the right hand side, making a tray to hold nails, screws, hammer, pliers, and other small tools used in fastening the plank, thus avoiding the necessity of stooping over the work, and also keeping the tools off the floor.

A framework of some description is always necessary to support the boat or vessel in building. If a ship or yacht, the keel is laid on blocks a short distance apart, but in boat work, the "stocks," as they are called, are usually a plank set on edge, at such a height above the floor as will bring

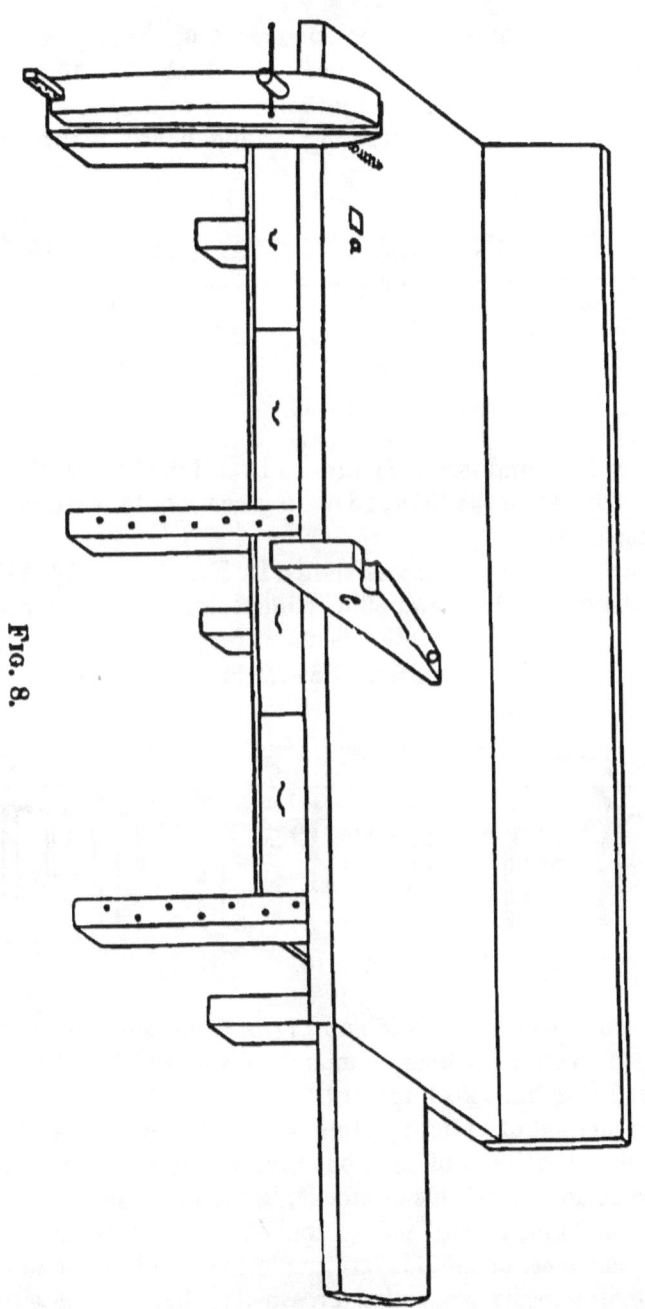

Fig. 8.

the boat in a convenient position (Fig. 9). The piece (*a*) is a common pine or spruce board, 1in. thick, 8 or 10in. wide and 13ft. long, the upper edge being cut to the rocker of the keel, as taken from the drawing. This board is supported

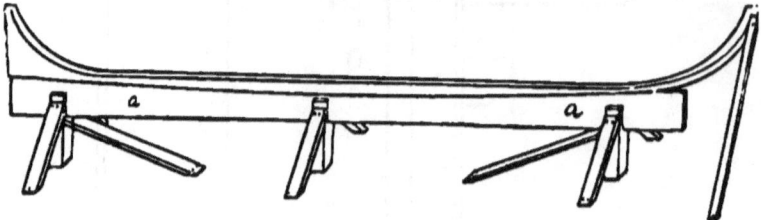

FIG. 9.

on three legs and securely braced in all directions, the top being 20in. from the floor, so as to give room to work on the garboards.

Another style of stocks is shown in Fig. 10, a table being built about 13ft. long and 30in. wide, somewhat like a canoe in breadth; the top, which is 20in. from the floor, is perfectly level. A line is drawn down the center, while across

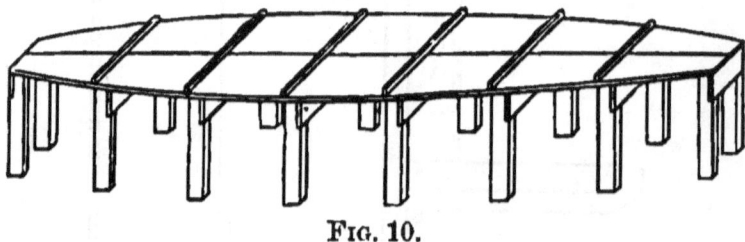

FIG. 10.

the board, battens, 1in. wide and 1½in. deep, are nailed, 2ft. apart, to each of which a mould is screwed, the boat, of course, being built keel upward.

This method of building (similar to that employed for shell boats) is the easiest and best, but involves more labor in the construction of the table or stocks; however, if several boats are to be built, it will pay to make a strong level table, as when once a set of moulds are made and each fitted to the screw-holes in its respective crosspiece, they may be set in place in a few minutes with every certainty that they are correctly placed, and that they must remain so, while the table makes a convenient place to lay tools.

Finally a steam box of some kind is necessary, its size depending on the work to be done. Usually all the steaming required for a canoe is the timbers, perhaps $\frac{1}{4}$ or $\frac{5}{16}$in. thick, which may be done with care in a trough of boiling water, but if anything larger is to be bent, a kettle, holding a couple of pails of water, should be arranged over a stove, or roughly bricked in if out of doors, a top of 2in. plank being fitted closely to it with a pipe leading from the top to the steambox, which is of 1¼ or even 1in. boards, and may be 8x6in. inside and 7ft. long, supported on trestles or legs near the kettle, and fitted on one end with a hinged door to close tightly, or the end may be closed with a bundle of rags.

(See Plate XV. and page 129.)

BUILDING.

THE drawing of the boat being completed, the moulds made from it and the bench and stocks being ready as previously described, the first step in the actual work of building, is the shaping of the keel. If the boat has no centerboard trunk, the keel is made of the same siding or thickness as the stem and stern, for its entire length, its depth below the rabbet being taken from the drawing and ¼in., the thickness of the plank, added. The keel may be made 1¼in. deep, the extra depth, if more is required, being made up by a false keel screwed to it, which may be removed for shoal water, as shown in Fig. 11. In selecting the wood for the keel and keel batten, the layers should lie horizontally, as shown.

If for a centerboard, either of the usual form, or one of the patented varieties requiring a trunk, a flat keel must be used as shown in the plate, which represents the cross section of a flat keel and centerboard trunk. The width, for the length of the trunk, will be 3½in. on top, tapering to the size of the stem and stern at its ends, the depth or thickness of the keel being uniform, ¾in. to 1in. throughout its length.

With the edge keel, a keel batten is necessary, as shown in the cross section. This will be ¼in. thick, and 1in. wider than the keel, to which it is nailed, thus overlapping the latter ½in. on each side, forming a rabbet for the garboards. If the flat keel is used, the rabbet is cut directly on the keel.

The stem is next sawed out from a hackmatack knee, and planed up ⅞ or 1in. thick, for an ordinary canoe, and the fore edge, rabbet and bearding lines marked on it, using the moulds made for each.

The rabbet line of a boat, marked *a* in the drawing, is the

AMATEUR CANOE BUILDING. 41

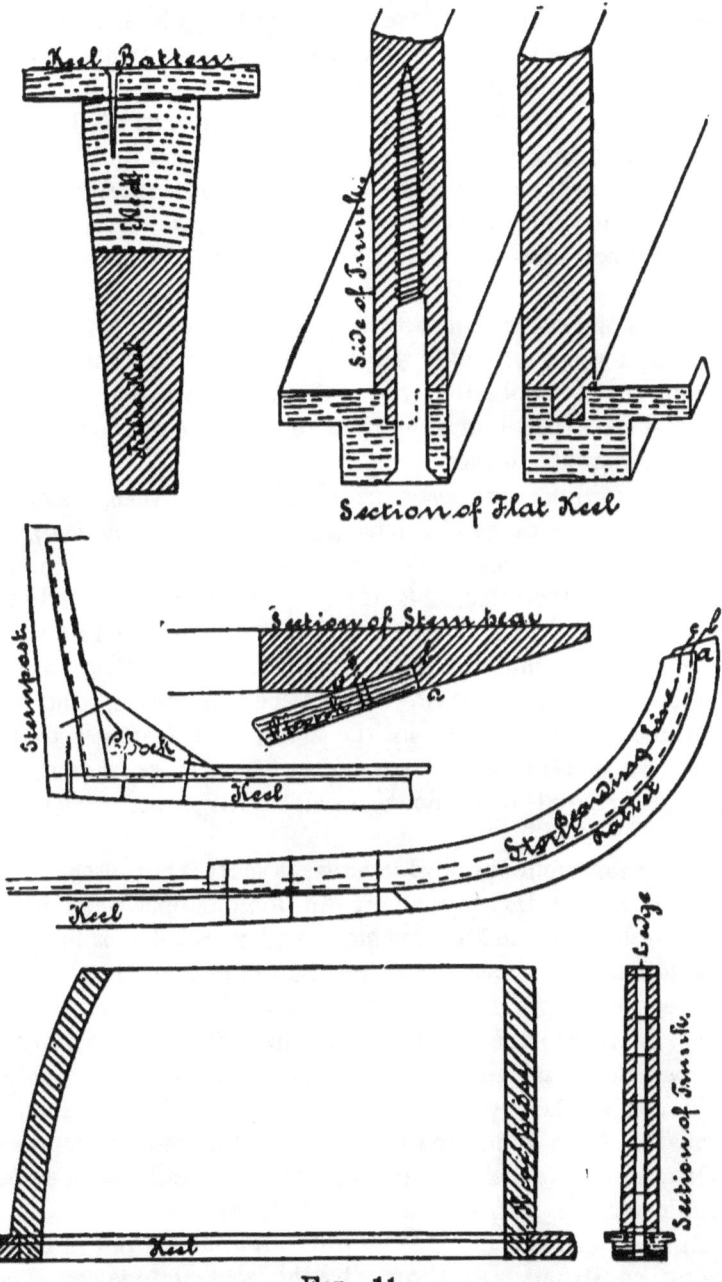

Fig. 11.

line where the outer surface of the skin or planking joins the surface of the stem, stern, and keel; the inner or back rabbet, *b*, shown by the dotted line, is the line along which the inner side of the plank joins the lower edge or ends of the same, and the bearding line, *c*, shown by a broken line, is where the inner surface of the skin joins the deadwoods, keel, stem and stern. The back rabbet is found by squaring in from the rabbet line, a distance equal to the thickness of the plank.

After the rabbet and bearding lines are laid off, the rabbet is cut, a piece of wood $\frac{1}{4}$in. thick and several inches long being used, applied to the rabbet as the cutting progresses to test its depth and shape. The rabbet is not cut quite to its full depth at present.

The sternpost in most canoes is made of a knee, the rabbet being curved as at the bow (see drawing of the Dot), but there is no good reason for so doing, unless the rake of the sternpost is excessive, as is now seldom the case, and a better plan is to make the sternpost of a straight piece, as shown, the rabbet forming a right angle or a little more, at the junction of keel and post. This piece is planed up, the rabbet marked and cut, as in the stem, and fastened to the keel by a 2½in. screw passing up into it, as shown, and further secured by a chock of oak nailed or screwed in the angle.

To fasten stem and keel together, a scarf is cut of the shape shown in the drawing, about 8in. long, copper nails being driven through the keel and stem, and rivetted over burrs on the top of the former. The keel batten is now nailed on top of keel, butting against the stem forward and the chock aft. The bearding line is drawn in where it has been omitted across the scarf forward and chock aft, and the rabbet trimmed at these points and the frame laid on the large drawing, from which the water line is marked on stem and stern, and the positions of moulds, bulkheads, mast steps, trunk, etc., on both top and bottom of keel.

If a centerboard trunk is required, it must be put in now; being constructed as shown by the sectional views. The head ledges, forming the ends of the trunk, are of oak, 1¼in. wide and as thick as the slot or opening, ⅜in. for a thin iron board, and ¾ to 1in. for a heavy iron or a wooden one. The slot is first cut, 1¼in. longer at each end than the required opening, then a groove, ¼in. wide and deep, is ploughed on each side of it for its entire length.

The head ledges are now fitted in place, projecting over the keel ¼in. fore and aft, to allow for caulking, and fastened by a copper rivet through the keel and lower end of each to keep the keel from splitting. The sides of the case, of dry pine, are ⅜in. thick on the lower edges, each of which has a tongue on it to fit the grooves in keel, and ¼in. on upper edges. A thread of cotton lamp wick is laid in the grooves, the inner surface of the sides, as well as their lower edges, the keel and the head ledges are well painted, and they are put in place and driven into the grooves. Before the paint is hard the sides are rivetted to the head ledges with 2in. copper nails, and brass screws 3¼in. long, spaced 6in. apart, are put through the keel up into the sides, the holes for them being very carefully bored and countersunk into the keel. If the board is hung on a bolt, the hole for it must now be bored, as it cannot be done later.

The moulds must now be fitted to their places, a small piece being cut out of each to admit that part of the keel and keelson inside of the bearding line, after which, if the boat is to be built with the keel down, the frame is placed in position on the stocks, secured by a few nails driven through the keel into the latter (which will be drawn and the holes plugged when the boat is ready to turn over), the stem and stern are plumbed with a plumb-line and fastened by shores from the floor or roof, the moulds put in position, adjusted by a center line from stem to stern, and also shored firmly.

If the latter method of building is followed, the moulds are screwed to the table, the frame laid on them, and all firmly shored from floor to ceiling. Now a ribband one-half inch square is nailed along on each side, at the height of the deck, being fastened to the stem, stern and the moulds, and the positions of the bulkheads and ribs are squared up or down on to them.

To prevent any leakage through the scarfs, stopwaters are next put in. These are small plugs of dry pine, the holes for which are bored where the seam or joint crosses the rabbet. They should be bored between the inner and outer rabbet lines, Fig. 12, so as to be covered by the caulking, if in a large boat, or by the edge of the plank where the seam is not caulked, as in a canoe. This should be done at all scarfs, or where water is liable to follow a seam.

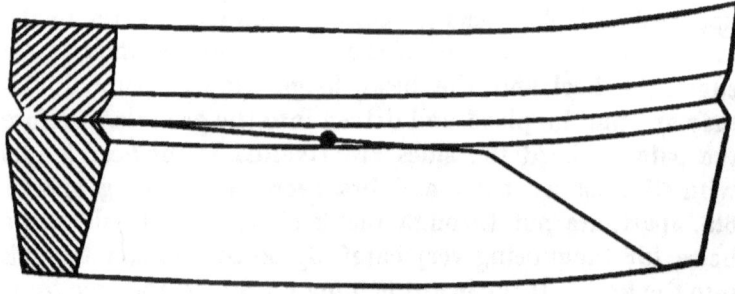

Fig. 12.

The rabbet is now completed by trimming it out with a sharp chisel, using as a guide, a strip 1x¼in. and long enough to cross at least two moulds. This is held down across the moulds, one end being applied to the rabbet, and the wood cut away until the surface of the strip and the outside of stem and stern coincide.

The positions of the ribs are now laid off, as shown in Fig. 13, which represents the fore end of a canoe, set up on a building table or bench. The distance apart of the ribs will be 5in., with an intermediate rivet through each lap between every pair of timbers Beginning at station 7 the spaces of 5in. are laid off toward bow and stern to within a foot of each end, and marked on top and bottom of keel so as to be seen from inside or outside when the plank is on, and also squared down on the ribband.

Perhaps the most difficult part of boat building, certainly the most difficult to make plain to a novice, is the planking. In order to obtain both strength and durability, each piece must be put on in such a way that it will bring no strain on any one part, and will not itself be forced into an unnatural shape, to attain which ends, though it may be bent or twisted, it must not be "sprung" edgeways or in the direction of its breadth, or it can never be made to fit properly. Although strakes are sometimes "sprung on" by experienced builders, the amateur should not attempt it, as the chances are that the framework will be pulled out of shape.

Before commencing to plank, the beginner can obtain an idea of how the planks must lie by taking a piece of board as long as the boat, 4 or 5in. wide and ½in. thick, tacking the middle on moulds 6 and 8 at about the turn of the bilge, and then bending the plank until it lies on all the other moulds, but not forcing it edgeways to or from the keel. The ends of course will come up higher on bow and stern than the middle, and if the piece be laid in a similar manner along the keel they will also be higher. The garboard streak, or that next the keel, will be 4 to 5in. wide in most canoes; then marking off the width desired, 4¼in., for instance, on moulds 6 and 8, the board mentioned above, having one straight edge, is laid over the moulds, its straight edge 4¼in. from the keel and the ends bent down and tacked to each mould and the stem and stern, and a mark is made where the board crosses, showing the position of the upper edge of the garboard. By upper edge is meant the edge nearest the gunwale, in all cases, whether the boat is built keel up or otherwise. With some models it will be better to vary somewhat from this line, of which the builder must judge for himself, according to the circumstances of the case.

Next, to lay off the upper streak, we will take a width of 3¼in. at midships, 2in. at bow and 1¾in. at stern, marking off these distances (Fig. 13) from the upper edge of the streak already marked by a ribband, and putting a similar ribband through these three points, bending it fair and marking where it crosses each mould. There should be six streaks on each side, so there still remain four to be laid off; to do which, the distance from the lower edge of the upper streak to the upper edge of the garboard on bow, stern and each mould is divided into four equal parts, making the planks all the same width on any given mould, though of course the widths on one mould differ from those on another, as the planks taper toward the ends, the girths at bow and stern being much less than amidships.

The planks being laid off, the next operation is to get the shape of the garboard, to do which a "staff" is necessary. This is a piece of board four or five inches wide, one-quarter inch thick, and as long as the boat, several, having more or less curvature, being necessary for the different strakes. For accurate work, especially where there is no help at hand, it is best to have two short pieces, each about one foot longer than half the boat's length. One of these pieces is cut roughly to the shape of the forward rabbet and fastened in place with a screw clamp, or a small piece of wood with a nail through it called a hutchock (*l*) Fig. 13. It is then bent carefully over the moulds as far as it will reach, and fastened to each with a hutchock. The staff should be of uniform thickness and quality so as to bend fairly, and is best cut so as to lie in the rabbet, though it need not fit closely. A similar piece is now fitted aft, lapping some two feet over the former, and the two are nailed firmly together, so as to preserve their relative positions when removed from the moulds. As the fitting of the garboard depends mainly on the manner in which the spiling is taken, great care is needed to prevent the staff springing or buckling in applying it.

When it is properly adjusted a series of marks are made with the rule and pencil on the rabbet line on the frame, and also across the staff, about two inches apart where the line is curved, as at the stem, and four inches where it is straighter along the keel. These marks are to insure the compasses being set at the same points in taking the spiling, and in transferring from the staff to the plank afterward, as will be understood later.

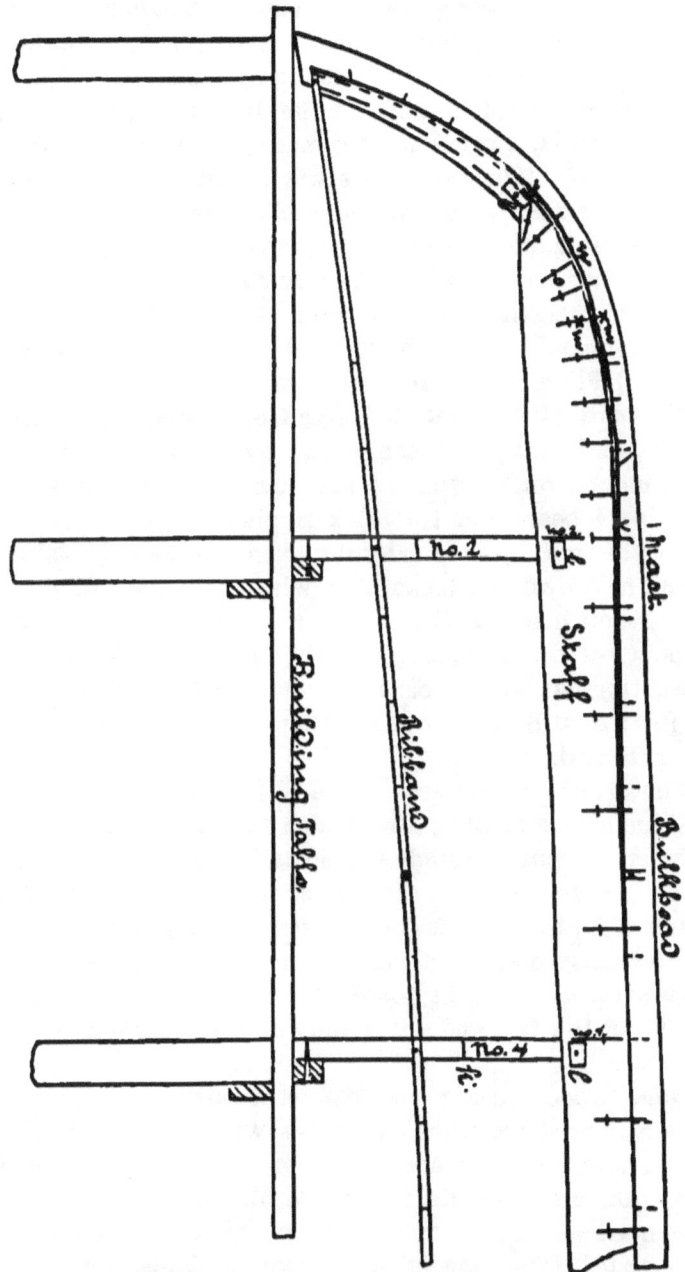

Now, with the compasses set to any convenient distance, usually from two to three inches, a circle is first swept on the staff, to reset them by if accidentally changed; then one point is applied to a mark on the rabbet line, as at n, and, with the other, a prick mark is made on the same line, at o on the staff. The compasses are applied in succession to each of the other points on the rabbet line and marks made on the staff, one line on the stem marked X X ($m\ m$) being called a sirmark, by which the plank is finally adjusted.

Before removing the staff from the moulds the position of each mould must be marked on it, as the breadths will be laid off afterward at each mould.

A board is now selected free from knots, sap or checks for the garboard. If it can be had planed to the thickness, ¼in., much trouble will be avoided, but where this is not possible, a board ¾ or 1in. thick is planed smooth on both sides, the staff is taken carefully from the moulds, laid on it and held by a few tacks, then with the compasses still set to the same distance, the measurements are reversed, placing a point of the compasses on the marks on the staff, and measuring out on the board. This operation, if accurately performed, will give the exact shape of the lower edge of the garboard.

The sirmark is now transferred to the board, and also the position of the moulds, after which the staff is removed and a batten is run through the spots, the curves on the ends being drawn in with the rabbet moulds. To lay off the upper edge, the breadths on the stem, stern and each mould, as previously marked off, are taken and transferred to the respective points on the board, an extra width of ⅜in. being added for the lap, and a line drawn through them with a batten.

Some woods, cedar and oak especially, will spring or change their shape when a strip is sawed off one edge, and if this happens, the shape may be so altered that it will be very difficult to make the plank fit. If a straight line is drawn down the center of the board before sawing, and then tested after one edge is sawn to shape, it will show whether the plank has sprung at all, and if it has. a strip should be sawn off the other edge, leaving the board still a little wider than the finished strake will be, and then the plank should be laid off anew from the staff, as in the first instance, after which the edges may be planed up, with little danger of further springing.

If the board is thick enough to make two strakes, gauge lines are now run around the edges $\frac{1}{4}$in. from each side, the piece is laid on the saw benches, one end wedged fast between the two upright pieces previously mentioned, and it is sawn through, using the rip saw held nearly vertical, a few inches being sawn from one edge, then the piece being turned over and sawn for a short distance from the other edge, this process being repeated until the sawing is finished, as the saw will certainly run if used entirely from one side

When the board is sawn in two. the pieces are each planed to thickness on the inside, after which the edges must be beveled to fit the rabbet. The best bevel for this purpose is made of two pieces of wood $\frac{3}{8}$in. wide and $1\frac{1}{2}$in. long, one piece, $\frac{1}{4}$in. thick, having a saw cut in one end, in which the other piece, $\frac{1}{16}$in. thick, is slipped. The bevel is applied to different points of the rabbet about 6in. apart in succession, and the angles transferred to the respective points on the strake, after which the entire edge is planed to correspond to these spots.

The second or broad strake will, of course, lap over the first, but at the ends the laps must diminish until the surface of both planks is flush with the stem at the rabbet. To secure this the adjoining surfaces of both are beveled off, beginning about 18in from each end and increasing in depth until about half is taken from each piece at the rabbet of stem and stern. This may be done with a rabbet plane or sharp chisel. The lower edge of the broad strake is left $\frac{1}{16}$in. thick, a rabbet being cut in the garboard to receive it, but the upper edge of the garboard is simply planed to a feather edge. Before cutting this rabbet the width of the lap, $\frac{3}{8}$in., should be marked with a scratch gauge on the outside of the garboard as a guide for setting the next plank.

All being ready, the garboard is now held in place, with the help of an assistant, each part of it being tried in the rabbet, to test the accuracy of the bevels. In doing this, the plank is not put in place for its entire length at once, but one end is tried, then the middle, and finally the other end.

The fitting being complete, the stopwaters in, and the hole bored for the centerboard bolt, if any; the garboard is fitted in place on the fore end, adjusted by the sirmark, the after part being held well up by an assistant, and one or more clamps are put on to hold it, then holes are bored and countersunk for the screws, which will be ⅝in. No. 5 brass, and the garboard is screwed fast as far as it lies in place.

In fastening such light plank, great care is needed to avoid splitting it; the pieces must be in contact before the screw or nail is put in, otherwise, if it is attempted to draw them together with the screws, the plank will usually split. Screws are only used at the extreme ends, where nails cannot be driven through and riveted, but along the keel the latter are put in. After the fore end is fastened, the plank is laid in place along the middle of the boat and nailed, every other nail being omitted to be put in after the timbers are in place, after which the stern is screwed fast.

If the operations described have been carried out correctly, the garboard should fit exactly without any further cutting, and the greatest care should be taken to do so, as if the strake does not fit at first, it is very difficult to make it do so by cutting it afterward. When both garboards are on, a spiling is taken for the broad strake; it is got out and put on in a similar manner, the staff, however, in this case being in one length. After the strake is in place and screwed at the fore end, it is fastened with clamps, and the positions of the nails, omitting all that will pass through the timbers, are marked off, using a thin batten bent around the boat, from the marks on the keel to those on the ribband, to insure the rows of nails being straight.

The nails for this work are of copper, ¾ or 1in. long. As the holes for them are bored, they may sometimes refuse to hold at first, in which case a block of soft wood, 1in. square is held inside the seam and the nail driven into it, the block being removed before riveting. It may sometimes be necessary to drive a nail through the plank into a mould, using a hutchock to hold the plank down, but this should be avoided if possible, as the hole will have to be plugged afterward.

To recapitulate, the process of preparing and placing a plank is as follows: First, to set the staff, mark it and take the spiling with the compasses, mark positions of moulds, plane both sides of board, remove staff, place it on board, nail it, spile off on the board, mark position of moulds on latter, remove staff, mark line of lower edge through the spots, lay off breadths at each mould on plank, leaving ⅛ extra for lap, line upper edge through these spots, saw out, plane up edges (if a thick plank, gauge edges, slit and plane insides), bevel edges, gauge upper edge on outside for lap cut rabbets at each end for next plank (on the bilge it will be necessary to bevel the upper edge of plank on outside for its entire length), put in place, clamp, screw fore end in rabbet, nail along lap, and cut and screw after end.

Where there is a quick turn to the bilge, it is best to use ¼in. stuff for each plank, hollowing the inside with a plane, and rounding the outside to fit the curve of the moulds. At the ends, where the laps are thinned down, tacks, ¼ and ⅜in. long, are used instead of nails.

The planking being completed, the canoe, if built with the keel up, is turned over on the stocks and shored in position, the keel being blocked to the proper rocker, then the ribs or timbers are sawed out of a piece of stave timber, ⅜x¼in., the upper corners are rounded off, and if not flexible enough to bend easily, they are put in the steam box or laid in boiling water.

The holes for the nails are now marked off by means of a wide, thin batten, which is bent into the bottom of the boat and adjusted to the mark on keel, and also so that it stands upright; then a mark is made where it crosses each lap, and a hole bored in the middle of the lap with a 1/16 in. German bit. When all the holes are bored, the ribs are taken one by one, bent over the knee and pressed down into the bottom of the boat, then the nails, which have previously been driven lightly into the holes, are driven up through the timber, using a set to hold on the top of latter alongside of the nail as it comes through. The lowest nail must always be driven first, then the others in succession from keel to gunwale.

As many ribs as possible should be put in before the moulds are removed, those alongside of the bulkheads, however, being omitted entirely. A nail must be put through the middle of the garboard and broad into each timber. After all are in, the boat is kept from spreading by means of cross spalls, pieces holding the gunwales together, and the moulds are removed; the blocks are then pulled off the ends of the nails, and the riveting up begins.

A copper burr or washer is slipped over a nail and driven home with a burr starter, an attendant outside holding the set on the head of the nail. When the burr is on, the end of the nail is cut off close to it, and the projecting part (about $\frac{1}{16}$in.) is headed with a few blows from a light riveting hammer, the tacks at the ends merely having their ends turned down. After the riveting is completed the gunwales are put on.

These were formerly put inside the boat, being jogged over the heads of the timbers, but a stronger and neater plan is to put them outside, making them of a hard wood, preferably mahogany. The deck is screwed to them, and they serve also as chafing battens, protecting the sides. They should be about 1¾in. wide at middle, 1¼ at fore and 1¼ at after ends, and ⅜in. thick. A rivet is put through the stem and both fore ends, and another through the stern, thus strengthening what was formerly one of the weakest points of a canoe. Nails are also driven through them and the upper streak and the head of each timber and riveted, making a much stiffer side than the old method. After the gunwales are in, the cross spalls may be shifted if necessary until the curves of both sides of the boat are perfectly fair and symmetrical.

The bulkhead timbers will be sawed from hackmatack knees $\frac{5}{8}$in. deep and $\frac{1}{2}$in. wide. They must be fitted accurately to place in order to make a water-tight joint, to do which, a piece of thin board is cut to fit closely, the timbers being marked from it. After the timbers are fitted as tightly as possible by this means, a little dark paint is laid on where the timber will come, the latter is put in place and pressed down, with a slight fore and aft movement, and on removing it, the points where it touches will be marked with paint.

These are cut away slightly, the piece replaced, and the operation repeated until the paint shows on the entire surface of the timber; it is then painted with thick white lead, pressed into place, and fastened by screws or nails through the planks at each lap and also in the middle of each strake, or if a wide strake, with two nails.

The bulkheads will be of white pine, $\frac{3}{8}$in. thick; they are placed on that side of the timbers nearest the end of the boat, and are riveted to them. A door is sometimes cut in the bulkhead to give access to the compartment in place of a deck hatch. These latter are to be avoided if possible, as they are never to be relied on as water-tight, and being exposed to rain and waves, are apt to wet all below, while a door in the bulkhead, even if not tight, is only exposed to water in case of a complete capsize.

It is still customary in many canoes to place the floor boards directly on the timbers, giving a little more space below deck, but allowing the water to cover the floor if there is the least leakage or a wave is shipped. A better plan, shown in Plate IV., is to raise the floor above the garboards from 1½ to 2in., according to the depth of the boat, thus giving space below for ballast if desired, and also keeping crew and stores dry, even though there is considerable water on board.

The floor is carried on ledges, z z, 1½in. deep at the middle by $\frac{3}{8}$in. wide, fitted closely to the planking, and secured by screws through the laps. Small limberholes should be cut in each piece to permit the free passage of water. These pieces also serve to strengthen the bottom of the canoe materially. The floor boards, $n\ n$, are in three widths, ⅜in. thick, of pine, the side pieces being screwed to the ledges, while the middle piece can be lifted out to stow ballast below. An oval hole in the latter piece, about under the knees of the crew, holds a sponge for bailing. The deck beams—of pine, spruce or hackmatack—are marked out from a beam mould, which is made from the large drawing. The amount of crown to be given to the deck must be decided on by the builder. From 3 to 3¼in. is not too much for a 30-in. boat, as the space below, for air and stowage, is much greater than with a flat deck; the boat will free herself from a wave quicker, and there is nothing to be said against it. Before putting in the deck beams the timbers must be cut off level with the gunwale, and the latter planed down until the sheer is perfectly fair from end to end, the beam mould being used at the same time as a guide by which to bevel the gunwales to suit the deck beams. The latter are spaced about as shown in the drawing, being fastened by a 2in. brass screw through gunwale and upper streak into each end. The beams will be 1in. deep and ½in. wide, except the partner beam that supports the mainmast, which will be 4in. wide, so as to take a 2¼in. hole for the mast tube, and the beams under the butts of the deck, which will be 1¼in. wide.

Canoe decks are sometimes laid in but two pieces, with one seam only, down the center, but while this makes a very handsome deck it is necessary to take off the entire half deck every time that repairs or alterations are to be made. It is often desirable to open one of the end compartments, and to do this quickly the decks are now very often laid in six or more pieces, one joint being over the forward bulkhead and one over the after one. At these points the beams are made 1¼in. wide and but ⅜in. deep, each piece of deck lapping ¼in. on the beam. After the beams are in, ridge pieces are fitted down the center of the deck fore and aft of the well. They are from 2 to 4in. wide, according to the size of the masts, and ⅜in. thick, being halved down into the deck beams and bulkheads and nailed to them. The holes for the mast tubes are now bored, the steps of oak are fitted and securely screwed or riveted to the keel and the mast tubes put in place. These are of copper or brass, the ends soldered up so that they are perfectly watertight. The upper ends are slightly flanged over the ridge pieces, with a little lamp wick and paint under the flange to make a tight joint. Plugs are sometimes put in the bulkheads to drain off any leakage, and the holes for them should be bored now, as low down as possible. The frame work of the well consists of two fore and aft pieces of spruce, $v\ v$, $\frac{3}{8}$x$1\frac{1}{4}$in. sprung partly to the shape of the well, the ends nailed to the deck beams and bulkhead, and also of two curved chocks, $r\ r$, at the forward end, completing the pointed form of the cockpit. The side decks are also supported by four knees, $y\ y$, on each side, sawn from oak $\frac{3}{8}$in. thick and screwed or riveted to the planking, a brass screw $1\frac{1}{4}$in. long passing through the gunwale into each, while the side pieces, $v\ v$, are screwed to the inner ends.

Before putting in the coaming, the decks, which will be of $\frac{1}{4}$in. mahogany or Spanish cedar, should be cut and fitted roughly to the outline of the well, the final fitting being done after the coamings are in. These should be of clear tough white oak, $\frac{1}{4}$in. thick. Their shape is taken by means of a thin staff sprung into the well, the upper and lower edges of the side pieces being marked on it. The pointed coamings now generally preferred are from 3 to $3\frac{1}{4}$in. high forward, sloping to $1\frac{1}{4}$in. amidships and aft, the after end being either round or square. The coamings are riveted to the side pieces and the after piece to the deck beam or bulkhead, a piece of $\frac{3}{8}$in. mahogany, g, being fitted in the angle forward, to strengthen it, and also to hold cleats and belaying pins.

The other fittings, described in the following chapter, such as side flaps, footgear, tabernacle, etc., are now put in, then the boat is turned over and the outside smoothed down, using fine sandpaper and a file on the nail heads; the stem band, of $\frac{5}{16}$in. half-round brass, is drilled and put on, the rudder braces are fitted and riveted fast, and sometimes bilge keels, which are strips of hard oak $\frac{3}{8}$in. square and about 4ft. long, are screwed to the bottom about over the second lap, protecting the boat in hauling up.

The outside of the boat and the inside of well has now a coat of raw linseed oil, and the inside of the compartments, the bottom, under the floor, and the deck frame, is painted with white lead and oil, sufficient black being added to make a lead color. Now, the bulkheads should be tested, to do which the boat is securely blocked up a short distance above the floor, and each bulkhead in turn filled with water, the leaks, if any, being carefully noted and marked. After the ends are tested, the water may be bailed into the middle of the boat, and the leaks there marked also. When these have been made tight, the decks may be laid, the pieces being first fitted, and then the under side of them being painted, and the edges of the gunwales, ridge pieces and bulkheads being also covered with thick paint or varnish. While this is fresh the pieces of deck are laid in place and fastened with $\frac{3}{8}$in. No. 5 brass screws, placed 3in. apart, along the gunwales, ridge pieces, deck beams, bulkheads and side pieces of the well. In all the older canoes the screw heads were countersunk and puttied over, but it is customary now only to screw them flush with the wood, allowing the head to show. If puttied over it is difficult to remove them, and the decks will be more or less defaced in clearing out the hard putty in order to do so.

After the deck is on, enough quarter-round beading of mahogany must be got out to go around the well, and also some half-round, to cover the seam down the center of the deck. These are nailed with half-inch brass or copper nails. The decks are next oiled, the mast plates, cleats, screw eyes, and other fittings screwed fast, the rudder, hatches, etc. completed, and all the outside of hull and inside of the well is varnished with some variety of wood filler, of which there are several in the market. This first coat is merely to fill the grain of the wood, and has no polish of its own. After it is thoroughly dry, a coat of spar composition should be given, and allowed full time to dry before using the boat.

CANOE FITTINGS.

WHILE the first requisite in a canoe is a properly-designed and constructed hull, there are a number of minor parts, generally summed up under the head of "Fittings," that are hardly less essential to safety, comfort and convenience, and which, with the sails and rigging, make up a complete craft. Perhaps a more correct term for these numerous details would be equipment, but the word fittings is generally used.

THE WELL.

This feature distinguishes the modern canoe from its savage progenitors, as, excepting the kayak, savage canoes are undecked, and its shape and position are important considerations. As a general rule, the smaller the well, the better; as less water can get below, there is more covered stowage room, and the boat is much stronger; but, on the other hand, there must be an opening long enough to permit sleeping, storing long spars below, giving access to hatches below deck, and, on occasion, taking a companion. The wells of the early Rob Roys were elliptical, 20in. wide and 32 to 36in. long, requiring no hatch, the coaming, 1in. high above deck, being bent in one piece, as in the drawing. This small well, resembling that of the kayak, was almost a necessity, as the boat was so low and wet in rough water.

A step in advance was the old Nautilus well, which was from 4ft. 8in. to 5ft. long, and 20in. wide, a length of 16in. being shut off by a movable bulkhead just abaft the crew's back; this portion being covered by a movable hatch, with a similar hatch at the forward end, leaving an opening of 2ft. or a little more for the crew. This well, with its ugly octagonal form, while a decided improvement in many ways, more than any other feature earned for the canoe the dismal epithet of coffin, once so frequently applied to it; besides which, owing to the number of pieces (eight) it gave no strength to the deck, and the joints soon opened and leaked, while the almost square end forward did not throw the water from the deck, but sent spray back over the crew.

In 1878 the Shadow canoe came out with an elliptical well 20in. by 5ft., covered by four hatches, so arranged as to close the well entirely in shipping the canoe; or by removing one or two hatches, making room for the crew when afloat. The first point was a decided advantage, but it was found in cruising that on a warm day the canoe became very hot below with hatches fitting closely around the canoeist, and when they were removed there was no room for them unless piled three high forward, and liable to be lost overboard.

At the same time the first Jersey Blue canoe appeared with a rectangular well 18in. by 5ft., 1ft. being abaft the crew, the coaming at sides of well extending over the forward deck and forming slides for a sliding hatch, which could be quickly pulled aft, covering as much of the well as desired, while a rubber apron, kept rolled up on top of the hatch, completed the covering. This arrangement answered the purpose of protection, but the square corners and sliding hatch were clumsy and heavy in appearance.

At the same time a canoe was built in Harlem having a pointed coaming forward, with a slight flare, the first of its kind, in America at least, and in 1880 the Sandy Hook and Jersey Blue canoes were fitted with pointed coamings, but not flaring, the first of the style now so common being put in the Dot in place of the Shadow well in 1881.

This form of well, shown in Plate IV., is in outline similar to a Gothic arch, and in addition the sides flare outward, throwing off all spray at the sides. The after end is made either round or square, the latter giving

AMATEUR CANOE BUILDING. 59

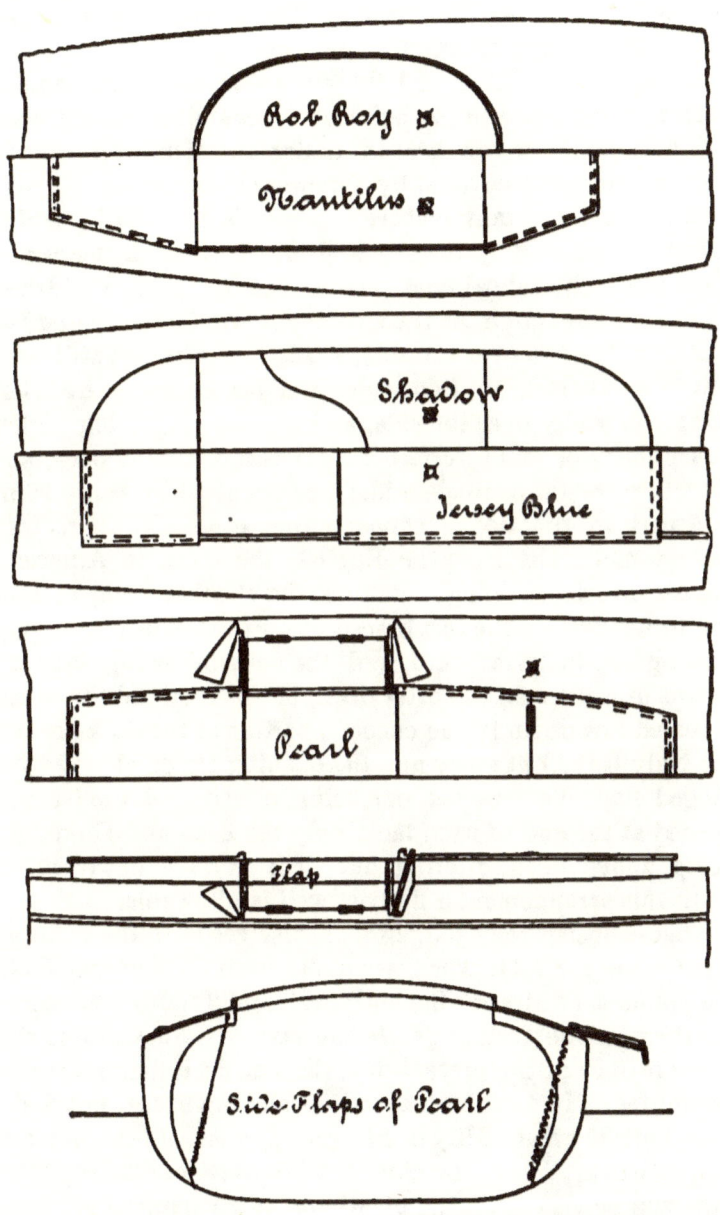

more room when two are carried. A chock of mahogany (*q*) in the drawing, is fitted in the angle, belaying pins or cleats being sometimes put on it. This form of coaming is well fitted to hold an apron, the fore end of which, being fitted to the point of the coaming, cannot wash off, and no spray can beat in under it. The well may be partly or entirely covered by hatches, as desired.

Another important feature in a well is its width, which must be regulated by the size and intended use of the canoe. In a narrow and shoal boat, such as the Rob Roy, a width of 18in. will be enough, as the side decks will be wider and less water will come over the side, while the crew can still lean out to windward, but in a wider and deeper boat there is less danger of water over the side, and the coaming being higher above the floor will interfere with the crew leaning over, and therefore should be made wider, the usual width being 20in.

American practice in canoe sailing, especially in racing, differs materially from the English; the crew, in America, almost invariably being seated on the weather deck, in sailing to windward, the feet braced under the lee deck, the body leaning well to windward, and the steering being done by means of a tiller on the after deck, but in England the crew is seated low down in the canoe, a portion of the deck abreast the body being cut away and the opening closed at will by a hinged flap, the weather one being closed and the lee one opened at the end of each tack, only the head and shoulders being above deck, offering but little surface to the wind. With this arrangement a narrow well is allowable.

That canoeists may judge for themselves as to the value of this feature for their work, we quote from the London *Field* the opinions of Messrs. Baden-Powell and Tredwen concerning them. The former gentleman says: "In describing the canoe fittings of the present day, the side deck flaps must not be omitted. In a sailing canoe it is all important, but I do not admit its great utility in a traveling canoe, at least not for general work. Where the chief work is to be lake sailing, side flaps will be very useful; but where much hauling out and jumping in and out is to be the order of the day, side flaps are utterly out of place. The side flap was first introduced in the Rob Roy in 1868, but did not appear in the next edition of that name. It has, however, now become a general favorite, and it is to be found in every sailing canoe. If fitted to the traveling canoe, the after end of the flap should be just forward of the backboard beam, and it should be strongly hinged at the outer edge; and, in short, strongly fitted in every way, as it is just about the place that one's hands lay hold of to raise the body in case of a sudden jump up or out. A broken, and perhaps lost overboard, flap would be a dangerous mishap to a canoe, if caught at the time in a breeze at a mile or two from land."

Mr. Tredwen, after describing some of the canoes that he has designed and built during the past fifteen years, continues: "It has already been observed that the flap side decks have not been fitted to all the Pearl canoes, and that where a canoe has been built with them, they have been subsequently discarded, and that the next canoe built without them has subsequently been altered by the addition of this contrivance. The result of this varied experience is to establish them as a very valuable adjunct to a cruising canoe if properly applied and fitted, otherwise they are better omitted. There are two essentials besides the flaps themselves, consisting of two sets of coamings around the openings cut in the deck. The first coamings are parallel and close to the cuts across the deck, and consequently at right angles with the ordinary well coamings, and are screwed securely to the deck, and their inboard ends butt on to the well coamings. They entirely prevent any leakage along the deck from forward or aft, into the openings of the flap side deck.

"The second set of coamings are placed transversely, hinged to the deck, and when raised their inboard ends fit closely against the beading or coaming of the hatch cover; and they are not intended to exclude leakage along the deck, but they serve as catches around which the mackintosh coat fits, to prevent any sea breaking into the well. The inboard ends must therefore project about half an inch above the hatch cover when they are raised. Many canoes have had these hinged coamings fitted without the fixed coamings, and without sufficient width to project above the hatch cover, and as they neither exclude water running back along the deck, nor provide a holdfast for the mackintosh, the whole contrivance has been condemned."

In this country the first step in this direction was in the Elfin, a New York canoe, which in 1878 had her coamings cut and hinged; the first real side flaps being put in the Sandy Hook in 1881, since which they have been tried in various canoes, but have not come into general use. Their construction is shown in the drawings.

In the Pearl canoe, the well, which is almost rectangular, is covered by a forward hatch in two parts, the after portion, extending to the body, being hinged to the forward part, so as to lie flat on it, when opened. On its after end is a beading, over which the skirt of the canoe jacket is drawn, this skirt also being held, by a rubber band run around its lower edge, to a similar beading on the after hatch, and to the hinged coamings described; the deck flaps opening inside the wide skirt, so that there is no entrance for water below. Where it is desired to close the canoe entirely, the well is covered by three or four hatches, fitting closely together, as shown in the drawing of the Shadow. These are held down by a bar running over them fore and aft, one end of which is inserted in an eyebolt at fore end of well, the other padlocking to a similar bolt aft.

APRONS.

In rainy weather or in rough water it is necessary to cover the well entirely, either by hatches or by an apron fitting closely around the body. The simplest form of apron, and one especially adapted to the pointed coaming, is a cover of cloth, cut to the shape of the coaming and turned down on the edges, to button over screw heads in the latter, near the deck. It also extends aft about 6in. over the hatch or deck immediately behind the back. A hole is cut for the body of the canoeist, and around the edge a piece (*a*) 6in. wide is stitched, so as to be drawn around the body. This piece is long enough to lap, as at (*b*), and button on one side. That portion of the apron abaft the body is held down by a cord (*c*) made fast to cleats or screweyes on deck, the apron not being buttoned to coaming abreast of the body.

A beam (*d*), to which the apron, just forward of the body, is nailed, keeps it arched so as to shed all water. If a forward hatch is used, the fore end of apron may be buttoned to it. In case of a capsize, the after part will pull from under the cord, and the canoeist is free, the apron remaining on the coaming. Instead of a buttonhole on the flap, a loop of light twine should be used, so as to break at once, if necessary.

*Another device is the telescopic apron devised by Mr. Farnham, which consists of a wire framework covered with oiled cloth. This framework is composed of several brass or German silver tubes (*e*), one sliding in another, as in a telescope, and also of carlins (*f*) of $\frac{1}{8}$-in. spring brass wire, soldered or brazed, as shown, to collars (*g*) on the tubes. The ends of these carlins are turned, as shown, to engage under the beading on the outer edge of the coaming, and are also bent into loops to avoid cutting the cloth. On the after end a piece of $\frac{3}{16}$ wire (*h*), bent to a curve, is brazed, being also brazed to the after carlin. This wire should extend 2in. aft of the sliding bulkhead to *i*. Forward of the well is a block screwed to the deck, and to it the first tube is pivoted by a universal joint, permitting a side motion to the framework, but holding it down forward, or it may be held by a strap, as shown. When the frame is drawn into position, the ends of the carlins, hooking under the beading, hold it down, and the curved ends of the piece (*h*) hook over blocks (*i*) on each side, keeping all in position.

The cover is of stout muslin, cut about 3in. larger each way than the coaming, so as to turn down, an elastic cord being run in the hem to draw it tight. Before sewing the cover to the frame, the ends of the carlins and all sharp corners or edges are covered with leather, so as to avoid

*See page 133.

cutting the cover. Extra strips are sewn on the lower side, under the carlins, to hold down the cover. For rough water an extra apron is used, being a short skirt, fitting under the arms, the lower edge gathered in by an elastic cord. An extra wire (k) is attached to the framework, forming a coaming on the after end of the apron, and a wooden coaming also runs across the after hatch. The lower edge of the skirt is drawn over these coamings, and also over two knobs (l) at the sides, the elastic holding all in place.

The apron on a Rob Roy or small canoe is sometimes held down by a strip of wood (m) on either side of the coaming, to which the apron is tacked, each strip having a flat brass hook (n) to hold it to the coaming, the forward end of apron being held down by a rubber cord passing around the fore end of well.

The material for an apron should be stout muslin, and after being cut and sewed it should be stretched tightly, well dampened, and coated with a mixture of turpentine one part, boiled linseed oil three parts, and raw oil six parts, laid on very thin, a second coat being given when the first is perfectly dry. To complete the covering of the well, either with hatches or aprons, a waterproof coat is necessary, made in the form of a loose shirt, opening about 6in. in front, the sleeves being gathered in at the wrists with elastic. The coat is just long enough to touch the floor when *seated*, and it should have a flounce outside, just under the arms, and long enough to fasten over the coamings, or hinged pieces of the side flaps, if the latter are used, the coat being full enough to allow them to be opened inside of it. To put on the coat it is rolled into a ring, slipped quickly over the head, the arms thrust into the sleeves, after which it may be adjusted at leisure. Care should be taken in putting it on, as an upset while entangled in it would be serious.

A seat of some kind is necessary in a canoe; it should be as low as possible, in order to keep the weight down, but still high enough to be comfortable when paddling. In a boat of 11in. or more depth the crew must sit several

AMATEUR CANOE BUILDING.

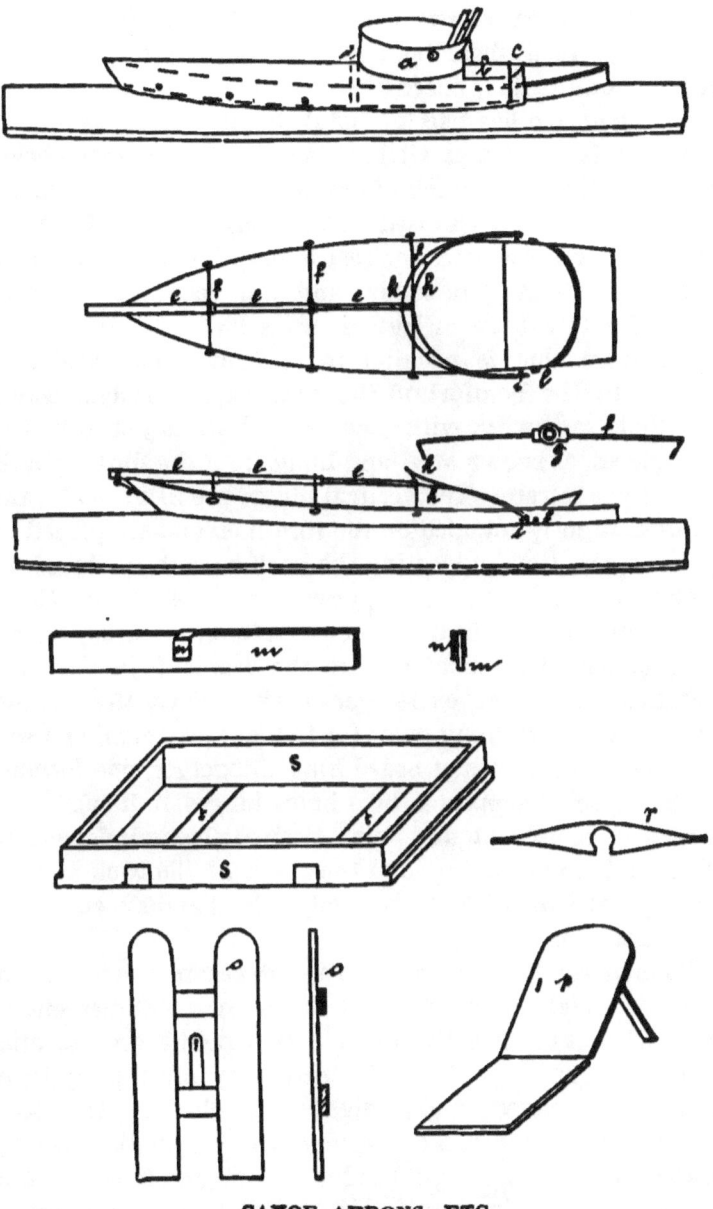

CANOE APRONS, ETC.

inches above the bottom to paddle comfortably, and in such a boat a high seat allows the body to lean further to windward; but in a shoal boat all that is necessary is a small cushion on the floor boards.

The tent, clothes bag or blankets may serve as a seat, though it is better that all bedding should be stowed below deck and out of the reach of any moisture. Some canoes are fitted with a seat of pressed wood, such as is used for chairseats, and in some cases the seat is simply a box without top or bottom, about 10in. square and 3in. deep, the top being covered with canvas, or leather straps.

A feature that is peculiar to the canoe, and that adds greatly to the comfort of the canoeist, is the backboard, usually a framework with two vertical strips joined by two crosspieces, as shown at o, and hung from the shifting bulkhead by a strap. The vertical pieces are $2\frac{1}{2}$in. wide and $\frac{3}{8}$ thick, slightly rounded on the fore side, and are placed $2\frac{1}{4}$ inches apart, thus supporting the back on either side of the backbone, and the crosspieces are rivetted to them. Sometimes a flat board, about 8x12in., is used, either with or without a cushion; but the frame is better. For paddling double, an extra beam is used across the cockpit, with a backboard hung on it for the forward man, or a seat is made of two pieces of board hinged together, one forming the back, being supported by a brace hinged to it (p). This back may be used at any point desired, being independent of the well and coaming, and the angle of the back may be changed at will, while it is easily folded and stowed away when not in use.

To increase the stowage room and to secure a better disposition of weights fore and aft, hatches are sometimes cut in the deck, but to be really valuable, two points are essential which have never yet been obtained; they must be quickly opened and closed, and airtight when closed. As good a method as any is to make a regular coaming to the opening in the deck $\frac{3}{4}$ to 1in. high, the hatch fitting on to this coaming with a beading projecting down, two thumb screws being used to secure it. Its water-tight qualities may be improved by a square of rubber cloth laid over the opening before putting on the hatch. This hatch is heavy and clumsy in appearance compared with hatches flush with the deck, but the latter always leak, and are never to be relied on.

In some cases where it may be desirable to get at the inside of the compartments occasionally for repairs, a hatch may be cut in the deck and covered with a piece of $\frac{1}{4}$in. mahogany decking, 1in. larger each way than the opening, and fastened by brass screws as the deck is, the laps being first painted. This piece will be airtight and yet can be removed and replaced in a few minutes when repairs are needed.

For transporting the canoe on shore a yoke is necessary, and may be made in several ways, the simplest form being that used for the guides' boats in the Adirondacks, a piece of wood (r) hollowed to fit over the shoulders and around the neck, the boat, bottom up of course, resting with one gunwale on each end of the yoke. Another form is a box (s) with no top or bottom, long enough to fit in the width of the well, and having two straps (t) across one side, which rest on the shoulders; the coaming of the boat resting on the ends of the box.

A plan lately devised by Mr. Farnham employs a frame of four pieces, which also serves in place of a sliding bulkhead. When used as a yoke, two straps are buckled across it and support it on the shoulders, the boat being inverted on it.

PADDLES.

The principal point of difference between a canoe and other boats, is the mode of propulsion, the paddle being held and supported by both hands, while in boats the oar or scull is *supported* on the boat, and its motion is *directed* by the hand. The former is the primitive mode, and even to-day the craft used by savage tribes are propelled almost entirely by paddles, the oar being used by civilized nations

The shape of the paddle differs greatly in various localities, but two forms only are known to modern canoeists, the single blade, shown in the center of the accompanying plate, and the double blade, various forms of which are also shown. The former, derived from the North American Indians, is about 5½ft. long, with a blade 5in. wide, and is made of maple, beech, or spruce. The upper end is fashioned so as to fit easily in the hand, the fingers being doubled over the top. The single paddle is used continuously on the same side of the boat, and its motion, in skilled hands, is noiseless.

The double paddle, the one best known in connection with modern canoes from the time of MacGregor, is derived directly from the Esquimau and his kayak. The length varies with the beam of the canoe, from 7 to 9ft., the former size being the one first used with the small canoes, but a gradual increase in length has been going on for some years, and of late many canoeists have adopted 9ft. instead of 8, as formerly, for boats of 30in. beam and over. Various patterns of paddles, as made by different builders, are shown in the plate, half of each paddle only being given. The blades vary in width from 6 to 7in., and in length from 18 to 20in.

Paddles of over 7ft. are usually cut in two and jointed, the joint consisting of two brass tubes, the larger one 5¼in. long and from $1\frac{5}{16}$ to $1\frac{7}{16}$ outside diameter; the smaller one 2⅜in. long, and fitting tightly inside the former. The short piece is sometimes fitted with a small pin, fitting notches in the longer piece, so that when the paddle is set, either with both blades in the same plane, or if paddling against the wind, the blades at right angles, no further motion is permitted in the joint; but this plan is not advisable, as when the joint sticks, as it often will, it is necessary to turn the parts to loosen them, which of course the pin prevents.

Tips of sheet brass or copper are put on the ends to preserve them from injury against stones and logs in pushing off. Pine or spruce are the best materials for paddles of this style. To prevent the water dripping down on the hands, rubber washers are used, or two round rubber bands on each end, about 2in. apart, will answer the same purpose. One

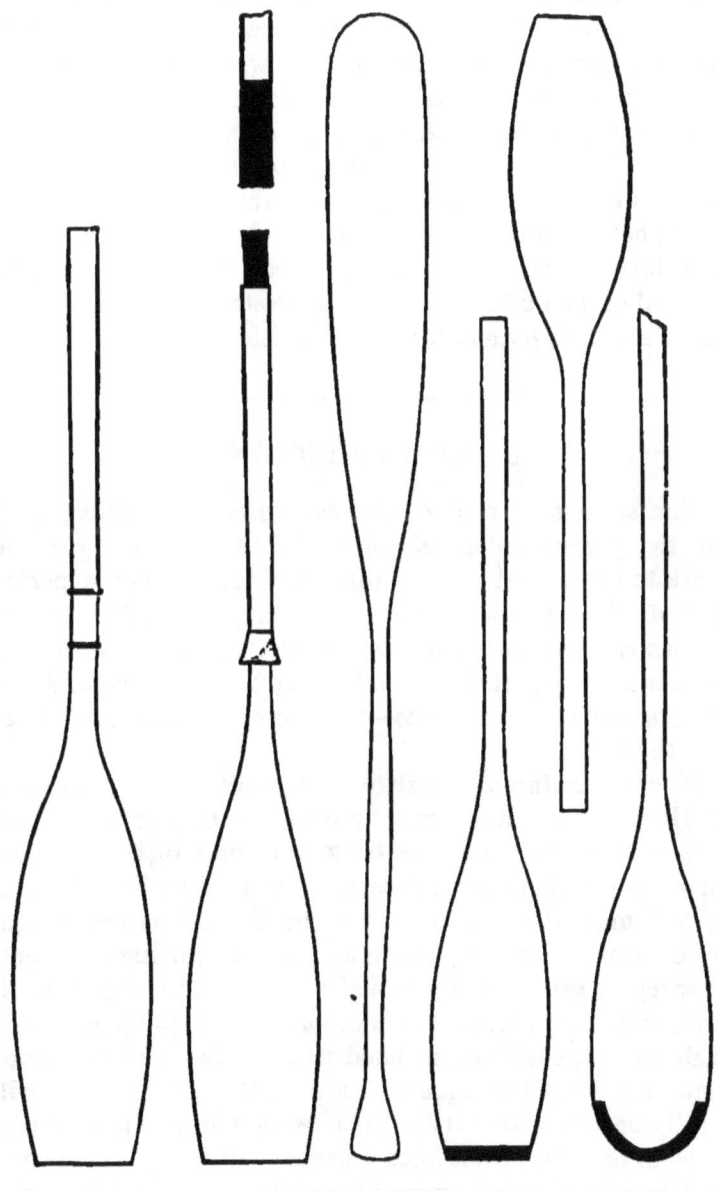

CANOE PADDLES.

half of the double paddle is sometimes used as a single blade, an extra piece, similar to the head shown on the double blade, being inserted in the ferrule; or when sailing, one half, lengthened out by a handle 18in. long, may be carried on deck, ready for any emergency, the other half being stowed below. The half paddle, in this case, is held with the blade under a cord stretched over the forward deck, the after end being held by a cord looped over a cleat abreast the body. For racing and light paddling, spoon blades are used, the general outline being the same as the straight blades, but the latter are stronger and better for cruising work.

SAILS AND RIGGING.

The success of a canoe as a sailing craft depends largely on the proportioning of the sails to the boat and the work to be done—on their proper fitting, and on the perfection of all the smaller details of the rigging. Almost every known rig has been tried on canoes, all but a few having been in time rejected, so that to-day but three types are at all popular with canoeists—the leg of mutton, the lateen, and the lug.

Before deciding on the shape of the sails, the first question is, How much sail to carry? a question only to be decided by a comparison with other boats and their rigs. Attempts have been made to formulate expressions by which the area of sail may be calculated when the dimensions and weight of the boat are known; but in a canoe the greatest elements in carrying sail are the personal qualities of the canoeist, his skill, activity, daring, prudence and good judgment; and their value is easily appreciated when on the same canoe one man can carry 100 square feet of sail, while another will hardly be safe with fifty. This being the case it is impossible to calculate what area a canoe will carry, but a comparison with similar boats will give the average cruising rig, the canoeist making such an addition to it as he considers will suit his individual wants.

Another uncertain element in carrying sail is the character of the water on which most of the work is done. If on a river or lake, among hills, where squalls are sudden and violent, the sails should be small, and the arrangements for furling and reefing them as complete and reliable as possible; if on open water, where the wind is strong but steady, a large sail may be carried, fitted with an ample reef for rough weather.

Whatever area be chosen, the almost universal practice with canoeists is to carry two sails. The cat rig, though simple, requires larger and heavier spars, a large boom and a high center of effort, and is more difficult to handle, as far as setting, furling and stowing sail, than the main and mizzen rig; and, on the other hand, a jib has been proved to be of little use, as it is difficult to set in a boat where the crew cannot go forward; a number of lines are needed, it requires constant attention, is useless when running, and of little benefit when doing its best. By having the bulk of the sail forward, it can be easily reached, is always in sight, draws well when running, and can be quickly spilled without losing the power of luffing, while the mizzen aft requires very little attention, luffs the boat promptly and keeps way on her, and even if neglected, can do little but bring her into the wind.

In a long, narrow boat like the canoe, the sail should be spread well fore and aft, long and low, rather than narrow and high, as the propelling power will be as great, and the heeling or capsizing power much less, and this end is best attained with the main and mizzen rig.

In order to obtain a proper balance of the sails, it is necessary that their common center or the point at which, if a force were applied, it would balance the pressure of the wind on the sails, which point is called the center of effort, should be nearly in the same vertical line with the center of lateral resistance of the hull, which latter is the point at which, if a string were attached, and the boat, with rudder amidships and centerboards down, were drawn sideways, it would advance at right angles to the string, neither bow nor stern being ahead. These points would be described in technical language as the common center of gravity of the sails, and the center of gravity of the immersed vertical longitudinal section, including rudder and centerboard.

The center of lateral resistance can be ascertained by drawing accurately to scale, on a piece of cardboard the outline of that portion of the hull below the waterline, including rudder, keel or board, taking it from the sheerplan, then cutting out the piece and balancing it on a fine needle stuck in a cork. The point on which it will balance is the center of lateral resistance.

To ascertain the center of effort, some calculation is necessary. A sail draft is first made showing the sails, masts, hull and center of lateral resistance, the scale being usually ¼ or ½in. to the foot for a canoe or small boat.

First, to determine the area of the sail, if triangular, a line is drawn from one angle perpendicular to the opposite side, or to that side produced. Then the area will be equal to one-half of the side multiplied by the distance from the side to the angle; for instance, in the triangle B C D in the first figure, which represents the calculations for a sail of 89 square feet, a line perpendicular to C D would not pass through B; so C D is produced to g then 12ft. 3in.x7ft. 6in. —91.87÷2—45.9ft., area of B C D. If the sail is not triangular it may be divided into several triangles, each being computed separately. The sail shown will first be divided by the line C D from throat to clew; the area of B C D has been ascertained to be 45.9ft., and similarly the area of A C D is 42.9, then the entire area will be 88.8ft. A shorter rule, and one that in most sails is sufficiently correct, is to multiply the distance A B by C D, and to take half of the product, but in a high, narrow sail, this would not answer, as in this case, where 16ft. 4in.x12ft. 3in.—200÷2—100ft., or an error of 11ft.

The area being known, the center of gravity of each triangle is next found by drawing a line from the middle of one side to the opposite angle, and laying off ⅓ of this line, as in the triangle, B C D, where half of C D is taken at a, a line,

a B, drawn, and ⅓ of it taken, giving the point d, the center of the triangle. The point c is found in a similar manner,

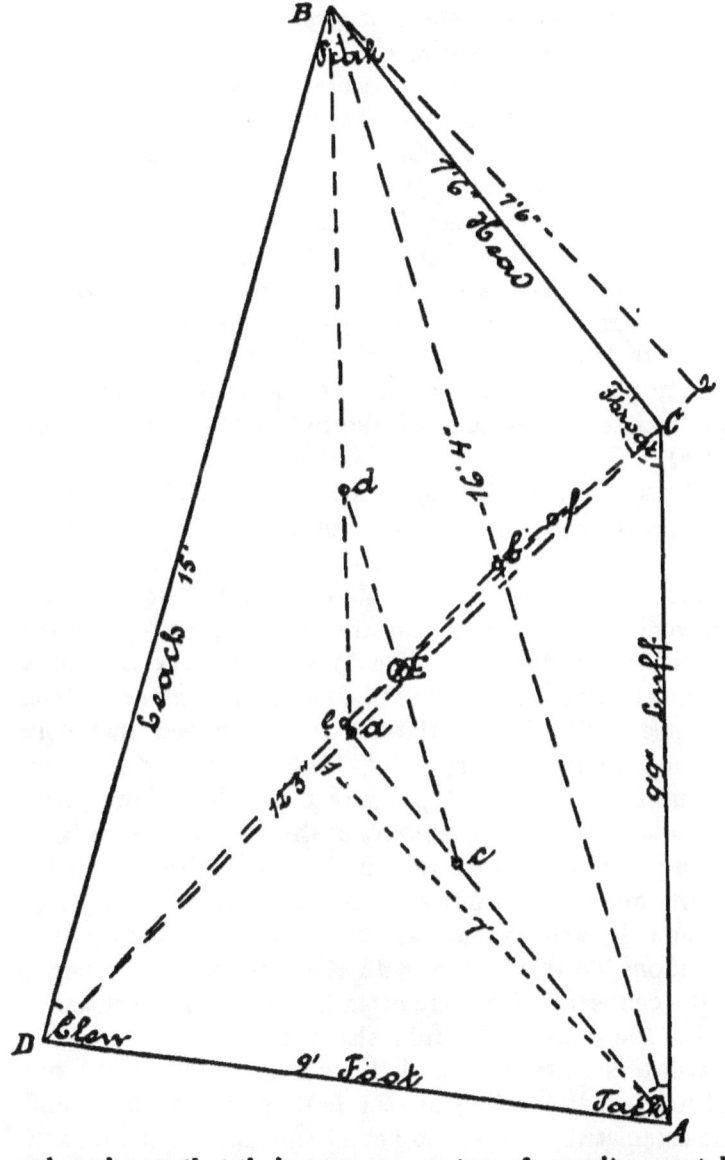

and we know that their common center of gravity must be on the line $c\,d$. Now, dividing the sail by a line, A B, into another set of triangles, A B C and A B D, we find their centers at e and f, and drawing the line $e\ f$, its intersection with $c\ d$ will be the center of gravity, and consequently center of effort, of the entire sail.

To determine the common center of two or more sails, a vertical line is drawn just ahead of the forward sail, and the distance of the center of each sail from this line is measured and multiplied by the area of the sail. In the drawing, showing two balance lugs of 45 and 20ft., the cruising rig. for a 14x30 canoe, these figures would be 40x5ft. 2in.=232, and 20x13ft. 7in.=273, or 505. Now, dividing this sum by the total area of the sails, or 65ft., we have $\frac{505}{65}$=7.77, or 7ft 9in., the distance of the center of effort from the vertical line. In this case, the center of effort of the sails and the center of lateral resistance of the hull will fall in the same vertical.

To be safe, a boat should always carry sufficient weather helm to luff easily, or in other words, when sailing on the wind, the leverage of the after sail should be enough to require that the helm be carried slightly on the weather side to prevent her coming up into the wind, then if it be left free she will luff instantly. To do this requires in theory that the center of effort should be aft of the center of lateral resistance, but in the calculations it is assumed that both sails and hull are plane surfaces, while in reality they are both curved and the wind pressure is distributed unequally over the sails; while the pressure of the wave on the lee bow, aided by a decrease of pressure under the lee quarter, tend to shove the boat to windward, independently of her sails, so that she will have a greater weather helm in any case than the calculations show, varying with the fullness of her bows, and the center of effort may often be placed some distance ahead of the center of lateral resistance.

It will be seen from this that such calculations are not absolutely exact, but they are the best guides we have, and if the calculated centers, and actual working in practice of different boats are recorded, a comparison will show what allowance is necessary in the case of a similar boat.

In planning a canoe's sails then, three things should be kept in view; to distribute the sail well fore and aft, keeping a low center of effort; to keep the latter about over the

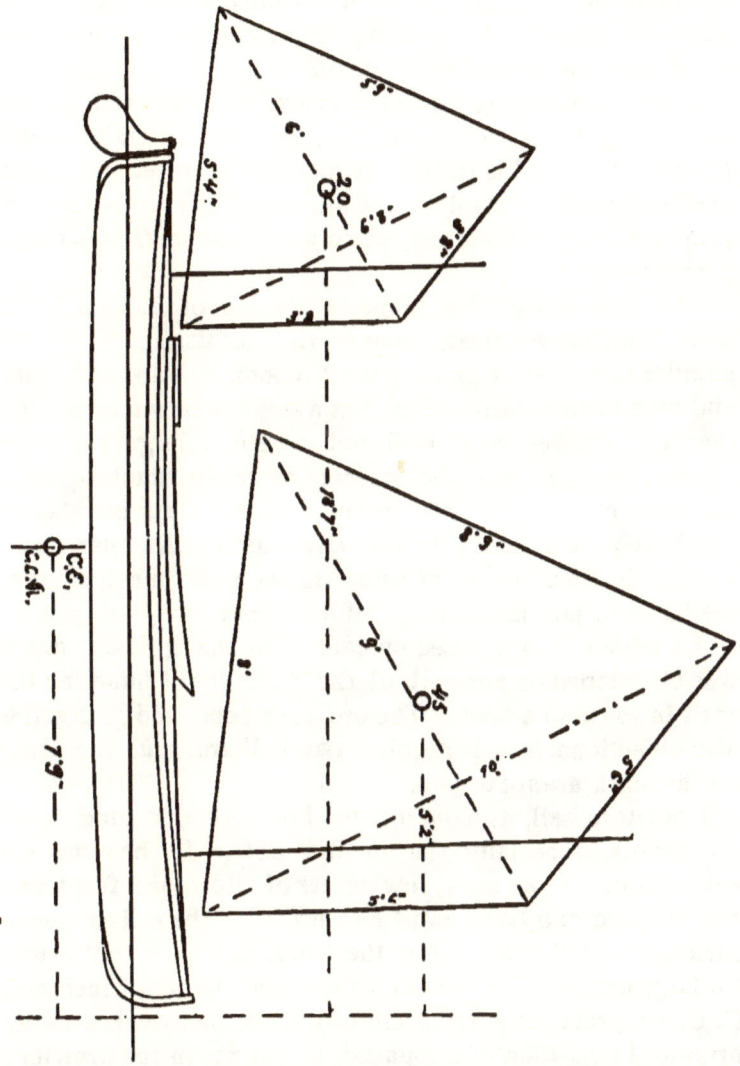

center of lateral resistance, and to keep as short a main boom as is consistent with the first point.

In order that a boat should sail equally well with her

board up or down, the center of the board should come under the center of lateral resistance, otherwise, if the board be forward and the boat balances with it lowered, on raising it, the center of lateral resistance at once moves aft, and the center of effort being unchanged, the greater leverage is forward, and the boat's head falls off.

If it is necessary to place the board well forward, it may be done by using a small mizzen, a reef being shaken out in it when the board is raised. A mainsail is sometimes rigged and tried with a cheaply made mizzen of any shape until the proper balance is obtained, when a suitable mizzen is rigged permanently.

The simplest rig for a canoe is the leg of mutton, or, as it is sometimes called, sharpie rig, consisting of two triangular sails, requiring only mast, boom, halliard and sheet, and on a narrow boat, where but a small area can be carried, they will answer very well, but where a large spread is needed, the spars must be so long as to be unmanageable; for instance, to spread 60 square feet, with an 8ft. boom, would require a mast 16ft. above the deck. Another disadvantage is the necessity of using rings on the mast, as they are liable to jam in hoisting and lowering.

A simple sail, once used on canoes, is the spritsail, but it was abandoned on account of the difficulty of handling the sprit in so small a boat. The ordinary boom and gaff sail is also objectionable as it requires two halliards and the rings on the mast, are apt to jam.

The lateen sail, as adapted by Lord Ross, is much used on canoes, especially the smaller ones. It has the advantages of a short mast, low center of effort, and few lines; but the yard and boom must be very long, the sail cannot be furled or reefed when before the wind, and it is not suited for large areas. The lateens introduced by the Cincinnati C. C. are practically leg of mutton sails, the yard peaking up into the position of a topmast, as shown in the drawing. The ordinary lateen rig consists of a triangular sail laced to a yard and boom, both spars being jointed together at the tack, and a pole mast with a spike several inches long on

AMATEUR CANOE BUILDING. 77

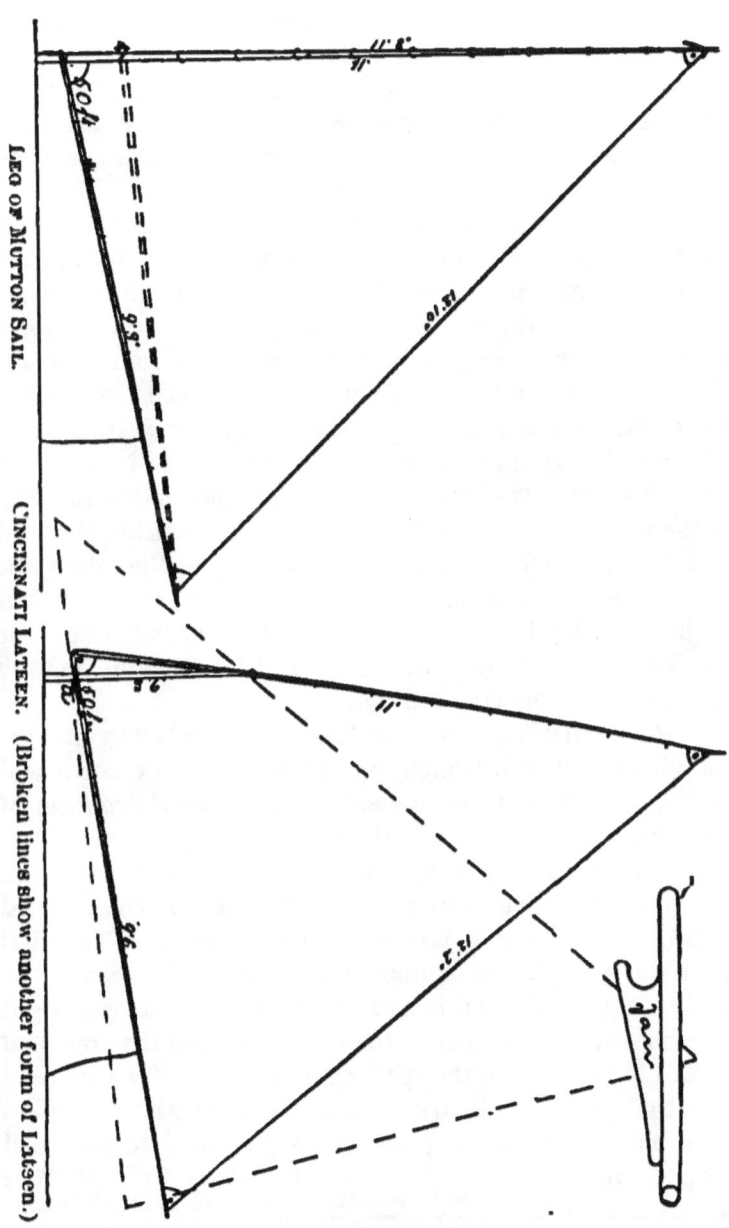

Leg of Mutton Sail.

Cincinnati Lateen. (Broken lines show another form of Lateen.)

top. A brass ring is lashed to the yard near its lower end, and a jaw (*a*) of wood or metal is fastened to the boom, a short distance from the forward end. In setting the sail, the yard is lifted until the ring can be hooked over the spike on the mast, then the boom is drawn back, lifting the yard, and the jaw is dropped in place around the mast, the operation being reversed in taking in sail.

The following method of reefing the lateen (see p. 83) was devised by Gen. Oliver, of the Mohican C. C. The fore end of the boom is fitted with a jaw (*b*) which encircles the mast when the sail is set, making a leg of mutton sail, while on the boom is a jaw (*a*). In reefing, the jaw (*b*) is removed from the mast, allowing the boom to come forward until *a* touches the mast, the slack of the sail being taken in by a reef line, *d d d*. One end of this line is made fast at the tack, it is then rove through grommets in the sail, and the other end made fast on the leach, the slack being taken in by hooking the cord over a screweye (*e*) on the boom forward, and another aft. Another similar plan dispenses with the jaw on the end of the boom, using instead a second jaw on the boom near the first, the shape of the sail being a little different, but the details of reef line, etc., the same.

Another sail devised by Gen. Oliver, and called by him the "Mohican" sail, is intended to combine the short boom and facility in reefing of the balance lug with the short mast of the lateen.*

Fig. A represents the sail set. The short mast with pin, and the spars toggled together of the Ross lateen, are used with the addition of a jaw at the end of boom. The sail is set in the usual lateen manner, and the spar, B, becomes virtually a high mast, and is treated as such. Four very light bamboo battens are put in the sail to increase the area, and the sail is attached to the spar, B, as far up as the ring, and from that point to a batten (*a*), and this batten is attached to B by halliard, *b*, which passes through block to foot and back to hand. The sail can be lowered by halliard (*b*) or taken off mast, A, in the usual manner of lateens.

The first reef is taken by letting go halliard and pulling in

* This sail is little used at present, and the name "Mohican" is applied solely to the settee sail described on page 159.

reef line (one being the continuation of the other) until batten touches boom. The Dot's reefing gear is used in this instance, and works admirably. The second reef is taken

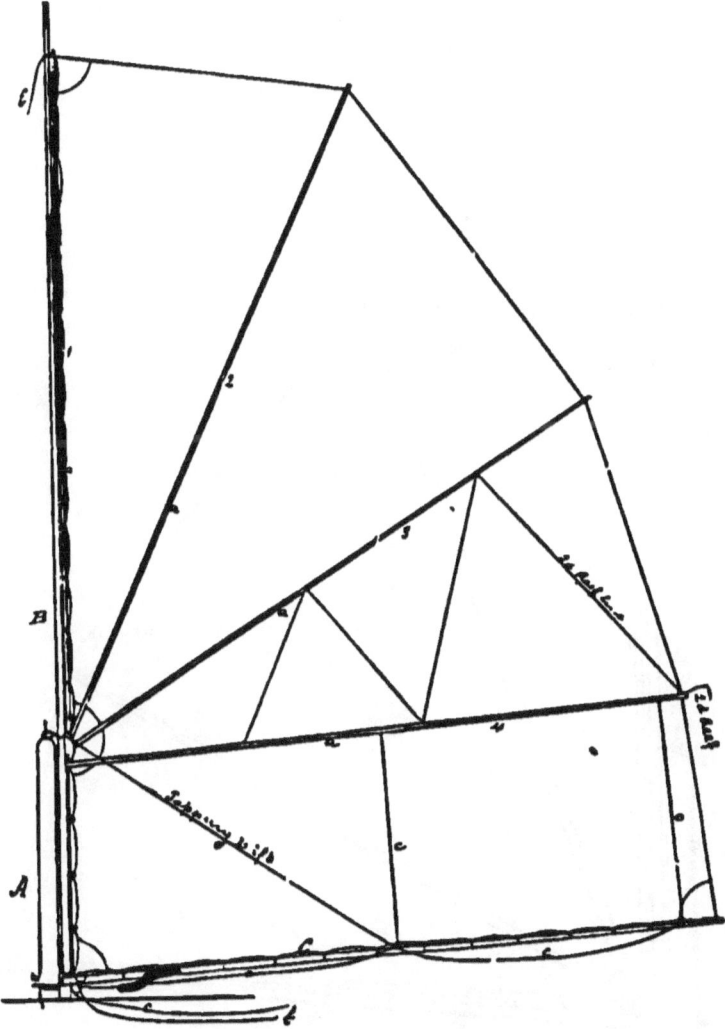

Fig. A.—"Mohican" Sail.

by unshipping boom C from mast A and hooking it again to A by the jaw. Batten No. 2 drops to No. 4, and the slack is taken up by reef line, as shown, and the sail becomes an ordinary lateen. The halliard and reef line may be made fast on boom, and should be so when sail is stowed away.

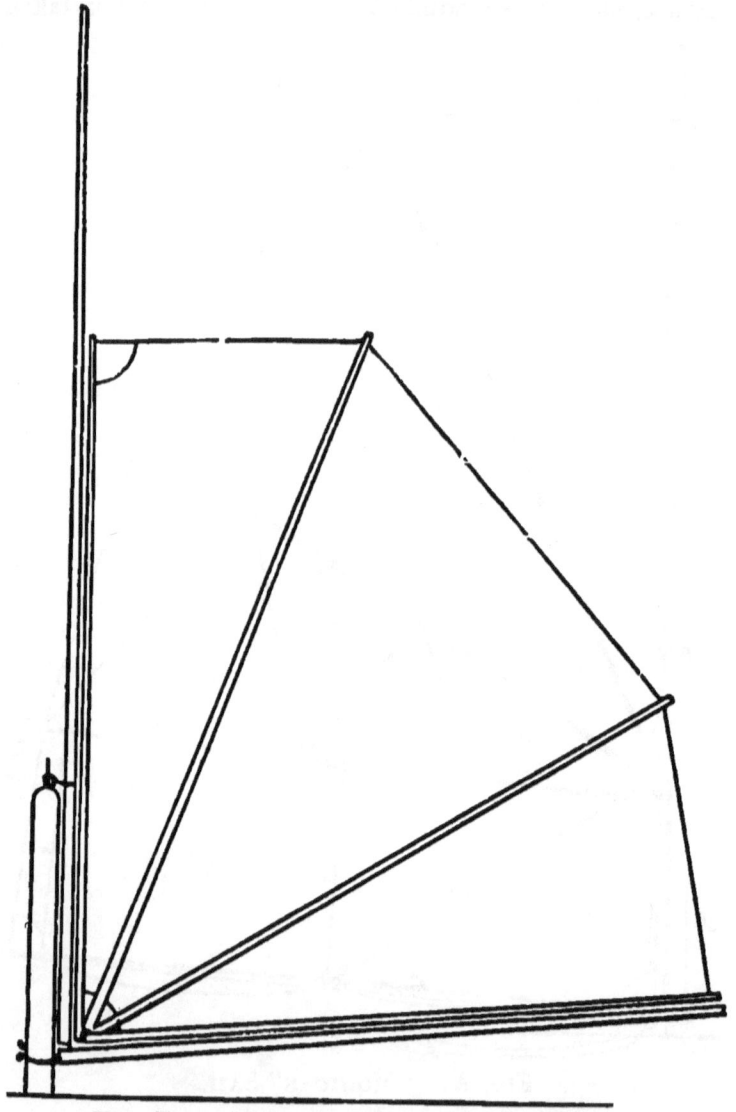

FIG. B.—"MOHICAN" SAIL SINGLE REEF.

This sail can be unshipped and stowed exactly as the lateen, and with the same advantages. It is always stowed on deck, made fast to side of coaming; and it has the reefing advantages of the balance lug, the short boom, and the height to catch light winds, with none of the disadvantages as to many ropes and high masts.

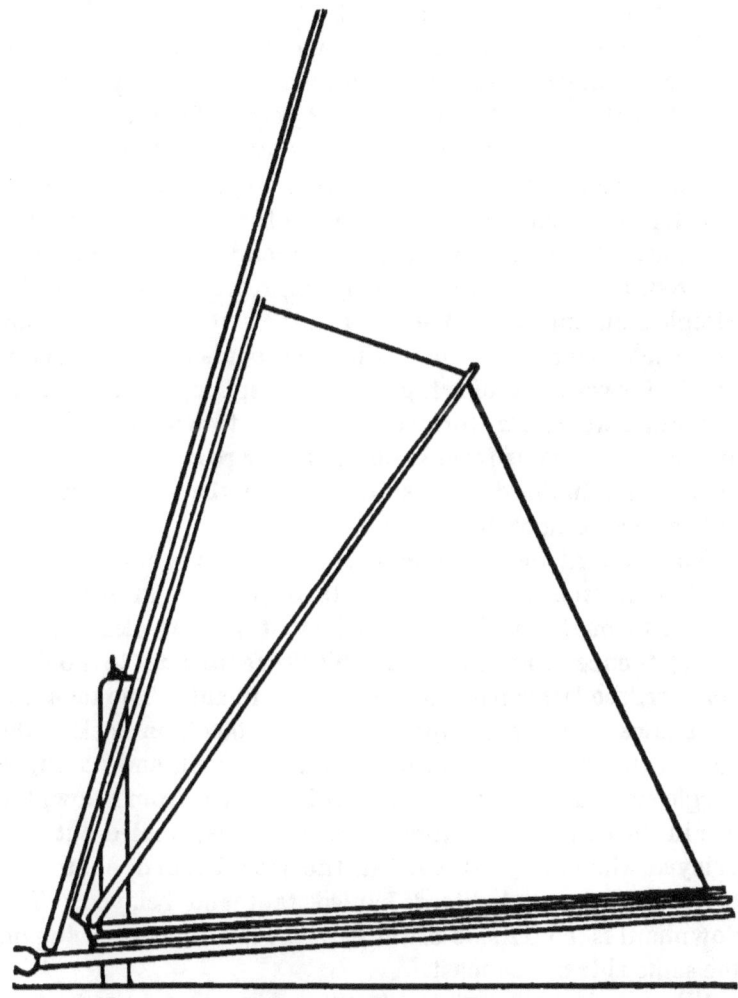

Fig. C.—"Mohican" Sail. Double Reef

A is the jaw; B, the spar or topmast; C, the boom.
Fig. B shows sail with one reef and Fig. C, with two. First reef can be taken in before the wind; second reef can

not, unless first reef is repeated with a parallel batten. In case leg o' mutton form of sail is used the area is much reduced, but all but No. 4 batten may be omitted, and the sail made fast to spar B by rings, and hoisted and lowered as in ordinary leg o' mutton sails.

The old sliding gunter is no longer used, as it was difficult either to hoist or lower the topmast with any pressure of wind on the sail. Within the past year (1888) the gunter rig has been revived, with better results, see page 192.

On a canoe, the nearer the sails approach a square. the shorter boom and yard they require for any given area, and the easier they are to handle and stow. All things considered, there is no sail so easily set, reefed or furled as the simple standing lug. The head of the sail is laced to a yard on which a ring b is lashed, while the foot is laced to a boom, in the forward end of which an eye is spliced. On the mast is a brass traveler a, formed of a ring to which a hook is brazed. An eye is formed on the upper part of the hook in which the halliard c is spliced, while the downhaul e is spliced to the hook itself.

The halliard and downhaul are sometimes in one piece, the lead being from eye in traveler through block at masthead, thence through double block at foot of mast to cleat on side deck; thence through double block again and to hook of traveler, the latter part forming a downhaul. The tack d is an endless line rove through a single block on deck at the foot of the mast and a screweye near the well, and having a toggle spliced into it. To set sail it is taken from below, the eye in the end of boom toggled to the tack, hauled out and belayed, then the yard is lifted, the ring hooked on to the traveler, and the halliard hauled taut and belayed. The downhaul is led outside of the sail, the latter always being on the same side of the mast.

Where a large area is to be carried, as in racing, the best sail is, beyond all question, the balance lug, a modification of the sails long in use in China, which was introduced to canoeists some fifteen years ago. In this sail a portion pro-

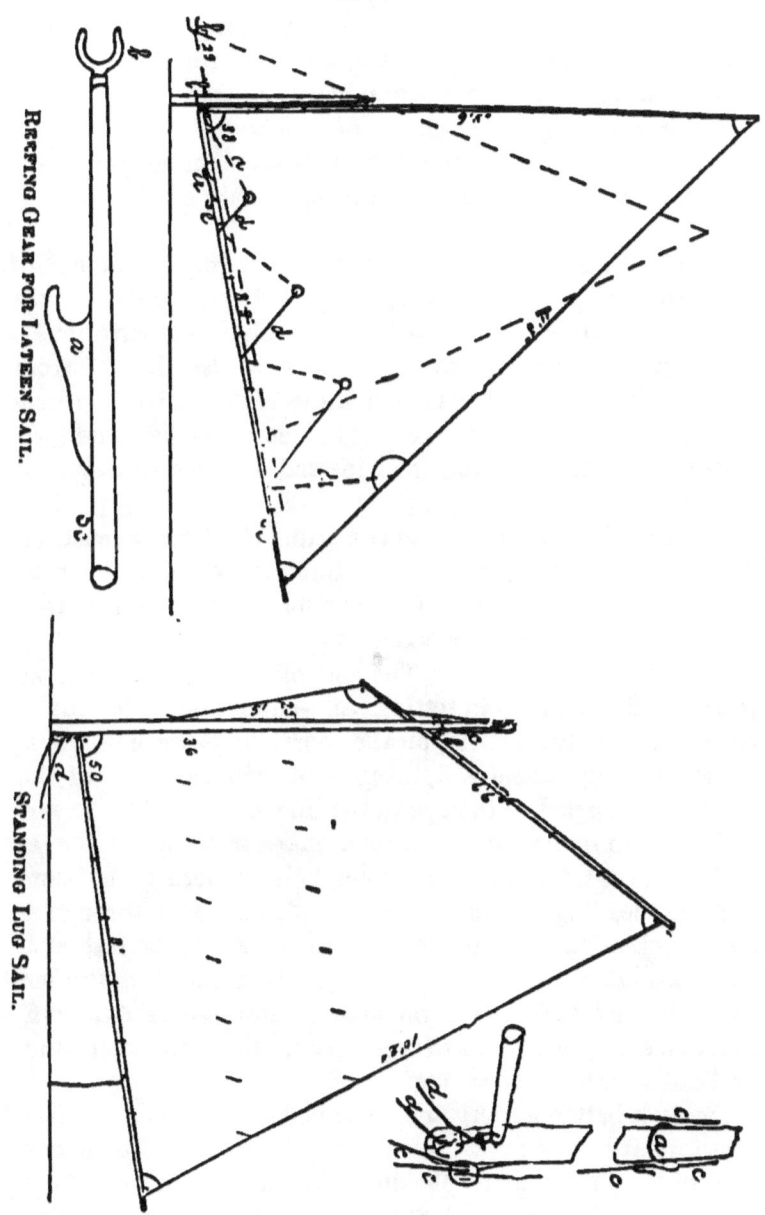

jects forward of the mast, greatly lessening the outboard weight when running free, as well as the length of the boom. The sail is spread on a yard and boom, as the standing lug, but is so hung that a portion hangs forward of the mast, about one-seventh to one-eighth of the boom being forward; thus, a sail of 7ft. on the foot will have no longer boom when running free than an ordinary sail of 6ft. on the foot.

To handle a large sail quickly and certainly a number of lines are needed, some of which may be dispensed with at the will of the skipper, but we will give all in the description.

One peculiarity of these sails, a feature also derived from the Chinese, is that they have a light batten sewn in a hem on the sail at every reef, keeping the sail very flat, and permitting the use of reefing gear instead of the ordinary reef-points.

The sail always remains on the same side of the mast, on either tack, being permanently hung there. On the yard just forward of the mast is a short piece of line (g), having an eye in one end, and a wooden toggle in the other, and abaft the mast is a thimble, k. The end of the halliard has an eye spliced in it, then in setting sail—supposing, as is usually the case, that the sail is on the port side—the halliard is passed through the eye k, around the *starboard* side of the mast, and toggled to the eye in the line g.

The boom is rigged in a similar manner, with thimble (k) and tack, the latter, about 5ft. long, being spliced to the boom at l, and leading around starboard side of mast through k and block m on deck, to cleat; or the tack may be fast at l, lead through a thimble lashed at starboard side of mast, then through eye k and to cleat on boom. In these sails the luff must be set up very taut to keep them flat, so the tack and halliard gear must be strong.

On each batten a short line (o), called a parrel, is made fast just forward of the mast, fastening with a toggle to an eye (p) on the batten abaft the mast, allowing such play as is necessary in lowering sail or reefing. These parrels confine the sail to the mast, keeping it flatter, and distributing its

AMATEUR CANOE BUILDING. 85

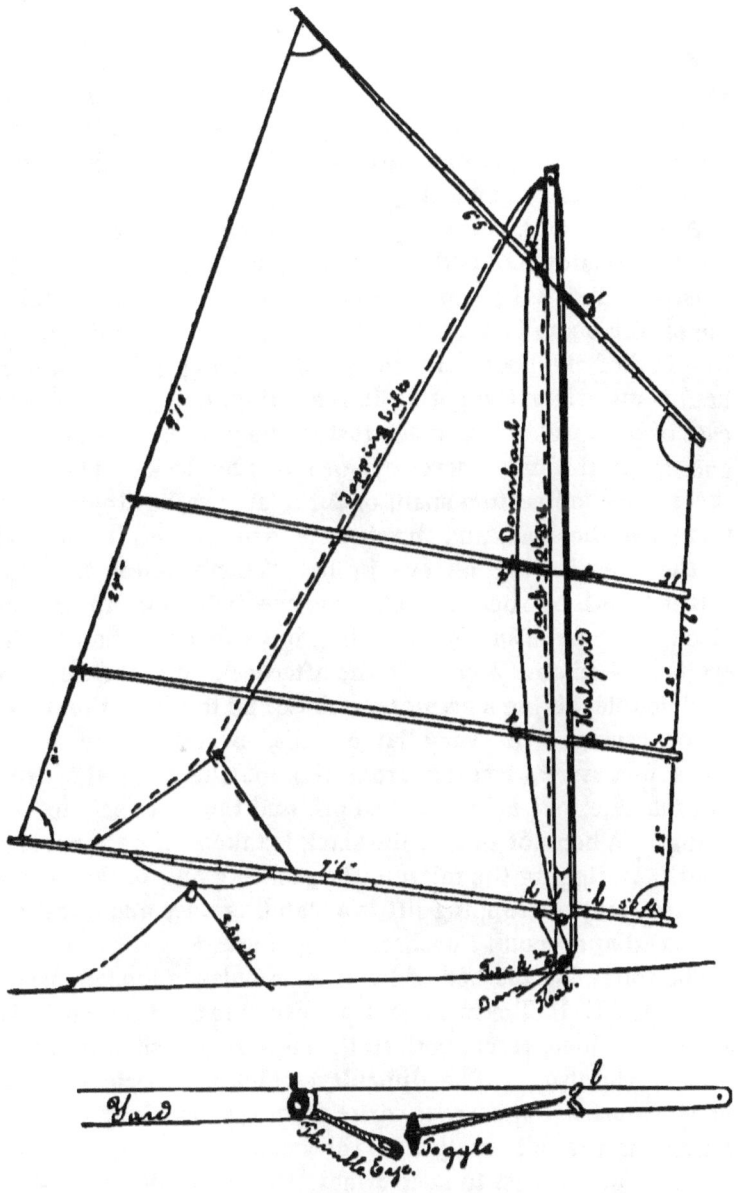

BALANCE LUG SAIL.

weight more uniformly over the entire length of the mast, thus easing the strain on the masthead.

A topping lift is usually fitted, being in two parts, one on each side of the sail. The lower ends are "crowsfeet," as shown, the main lines leading through a block at the mast head, and uniting in one part, which leads through a block at the deck and to a cleat.

Another line, t, called a jackstay, is made fast to the masthead, leads down outside of the sail, and is made fast to the mast just above the boom, or it may be led through a thimble on the boom to a cleat. Its purpose is to hold up the fore end of the boom in reefing and lowering sail. A downhaul is also rigged to gather in the sail quickly, especially in case of an upset. It is made fast to the yard near the eye, h, and leads through a screweye or block on deck. The main sheet is made fast to a span, or for a large sail a single block travels on the span, and the sheet is rove through it, one end of the latter having an eye in it. When running free, the entire length of sheet is used, the eye bringing up in the block and preventing it unreeving, but when closehauled the eye is hooked over a cleat on the afterdeck, and the sheet is used double, giving a greater purchase and taking in the slack.

For racing with very large sails, backstays are sometimes necessary, leading from the masthead to the deck on each side, one being slacked off, and the other set up, in jibing. When not in use, the slack is taken up by a rubber band. In rigging the mizzen, the jackstay and backstay are omitted, and the topping lift is a standing one, made fast to masthead and boom, the sheet being single. *

The following method of handling a balance lugsail, written by Mr. E. B. Tredwen, and published in the London *Field* some time since, refers both to the large racing sails, and to a cruising rig also: "The difficulty which is experienced by many canoeists, appears to arise from the needless labor of taking off the sail every time the canoe is housed. I have always found it best to keep a mast for each sail, a cruising mast and a racing mast, and the respective sails are never taken off their masts except for washing or repairing. Simi-

* A later and improved form of balance lug is described on page 225.

larly the mast which has been last used on the canoe is always put away with the canoe, either unstepping it, and laying it on deck, or lowering it (if a lowering mast be used) and letting it lie with a lashing to keep it in place.

If, however, the canoe must be left in the open, it is necessary to remove the mast and sail, which is very easily done. Having finished sailing and come alongside the boat house, the topping lift is let go and the after end of the boom comes on deck; then the tack must be slacked, or if the tack is a fixture, the jackstay must be slacked a few inches. The reeflines should next be gathered together, and stowed into a fold of the sail, the halliard and hauling part of the topping lift, similarly stowed in a fold on the opposite side of the sail, and the main sheet cast three or four times around all.

All the lines leading to the masthead (topping lift, halliards, etc.) should then be gathered to the mast about two feet above deck and a tyer put around. The after end of the sail can now be brought up to the mast and tied there, and the whole let run into a long bag and stowed away in the boat house.

When next going out for a sail, the mast is stepped, the tie of the boom end to the mast let go, and the sail brought down to the deck, the mainsheet cleared from around the sail, the topping lift set up, and the reeflines allowed to lie in the fore end of the well. The jackstay being set up, sail may be hoisted at once. The trouble when under way of reeving two reeflines through two screweyes, and knotting the ends for the sake of keeping them in their places, appears to be a detail scarcely worth discussion.

If the mast and sail are not taken off the boat at the end of the sail, there is not even the trouble of untying the knots in the ends of the lines. When my canoe sails have not been put away by a stranger, I can always get under sail in five minutes.

In a cruising sail there is no necessity for the tack to lead along the deck, or even along the boom. I have always cruised with a fixed tack about 6in. long, made fast to the lug of a triple pulley on the mast for the reefing gear to lead through.

The only occasion on which the tack need be started is in racing, when the wind is very light and the canoe is sailing between high banks. The tack may then be eased up until the yard is hoisted chock ablock, so as to get the sail as high as possible; at all other times a standing tack will do without any part on deck."

CENTERBOARDS.

In all decked canoes of classes A and B, which include probably two-thirds of the canoes used in America, sailing qualities have of late been considered as of even more importance than paddling, and the sailing powers of these boats have been developed to an extent never thought of by the first canoeists. Almost the first quality in a sailing boat is its lateral resistance, by reason of which it can be sailed to windward, and to secure enough in a canoe one of two things is necessary, a fixed keel, or a centerboard; the lee board being too clumsy a device to be of use in a canoe, though at one time occasionally used. An exception may be made here to the double leeboard used on the Canadian canoes, which may be handled on an open canoe, but will not answer for a decked one. There are a number of considerations on both sides of the question of keel vs. centerboard, and as no general rule is possible, we will notice the leading points on either side, leaving the canoeist to decide for himself after weighing them.

First—efficiency; the two are about equal as to lateral resistance and handling if the keel be rockered, otherwise the centerboard boat will turn more easily, and the double board is decidedly better than the keel when running free. Second—strength and weight; the keel boat will be stronger and lighter than any centerboard boat can be, but the latter can be built strong enough without being too heavy. Third—durability; the keel is not liable to accident and derangement as all boards are, and there is less danger of leakage, while the boat will stand more rough usage. Fourth—expense; the keel will cost usually from $15 to $25 less than a board of proper construction. Fifth—convenience; the keel boat gives more room inside, but will not stand upright on shore as the flat-bottomed centerboard will, which is a great disadvantage in landing, sleeping on shore and in packing stores aboard, and sometimes dangerous in running aground. On the other hand, a flat keel, as now built for centerboard, allows the canoe to rest in an upright position when on land, a very great convenience.

Whatever style of board may be adopted, to secure the best results it must be placed as nearly as possible in the proper position; but again the question of accommodation comes in. The best position for a board is, in most boats, with the center of its immersed portion a little forward of the center of lateral resistance of the hull and the center of effort of the sails; but in a canoe, in order to obtain room for sitting and sleeping, the board must be considerably forward of this if a trunk is required for it, and it may be moved forward without much harm, provided the aftersail is reduced in consequence. The only detriment to this plan would be that while the boat would balance properly on a wind with the board down, she would need a larger mizzen when in shoal water with board up. To avoid this disadvantage two plans are adopted, either to place the board well forward and add a second board aft, or to use a folding board that will not require a large trunk, and may be placed in any part of the boat. Of the latter class of boards there are several varieties, all of them patented.

The question of weight in a centerboard is also an important one. Most sailing canoes require some ballast, and in this form it can be carried lower than in any other way, as a drop of 18in. below the keel is allowed by the rules. The weight being so low down will make the boat much stiffer than inside ballast can, and its value will be found when running, as it will steady the boat greatly. The extra weight is of little account in handling, as the boards may be lifted out on landing so that the canoe and trunks will weigh no more than a canoe with fixed board. Several instances have occurred of canoes with heavy boards capsizing under racing sail until water poured into the well, but coming up safely and continuing.

The double board plan presents many advantages for a canoe, the center of the boat is entirely clear of trunk, lever or gear, leaving plenty of room for sleeping; with two boards, if properly worked, the boat may be handled to perfection in tacking, the canoe falling off quickly when the forward board is raised, and luffing when it is lowered and the after one raised, while in running free the after board steadies the boat greatly. The objection on the score of weight is but small, as both boards may be lifted out easily, when the weight of the two trunks is no more than that of most folding boards, while the boards themselves are ballast in its best shape. The smaller or after board will weigh from 7 to 12lbs., the forward one from 15 to 60lbs., as desired, or for light winds it may even be made of wood. These boards are also made so that a portion of the weight may be removed, as will be described further on.

The first point of importance in building a centerboard boat is the trunk for the board. In a boat of any size, a sloop or catboat, of 16ft. or upward, the trunk would be composed of two pieces of oak called bed pieces as long as the trunk, and for a small sailboat, 2x4in. placed on edge and bolted to the keel on each side of the slot, strips of canton flannel, painted with thick white lead, being laid between them and the keel. At each end of the slot are "headledges" also of oak, 2 or 3in. wide, in a fore and aft direction, and as thick as the width of the slot, which should be large enough to allow for the board swelling when wet. The slot being cut in the keel the headledges are driven into it at each end and a rivet put through each and the keel, then the bedpieces are put in place with the flannel between and bolted down to the keel, rivets being also driven through their ends and the headledges. The sides of the trunk are made of dry pine from 1 to 1¼ins. thick for a sailboat, riveted at the ends to the headledges, the seams between the sides and the bedpieces being caulked.

Such a construction is too heavy, and, besides, unnecessary in a light boat; the headledges (a a) are retained, but no bedpieces are put in. The headledges will be from $\frac{5}{8}$ to $\frac{3}{4}$ in. thick, according to the thickness of the board, and 1¼ in. wide, of spruce. They are set into the keel (b) as shown in Fig. 11, and also in plate on next page. The sides of the trunk are of well-seasoned and clear wood, usually white pine, although mahogany is more durable. A tongue is planed on the lower edge, ¼ in. wide and deep (see Fig. 11), and a corresponding groove is ploughed on each side of the slot. The sides are $\frac{3}{8}$ in. thick on lower edge, for a large board, but may be tapered down to $\frac{3}{8}$ in. at the top, as shown, to save unnecessary weight. Some care and neatness is required to make tight work; the sides are tongued on their lower edges, then fastened together, side by side, with a few small brads, and cut to the same shape; then the insides are painted, a strip of brass being first screwed to the inside of each to prevent wear, then they are carefully adjusted, with the headledges in place between them, and a few screws put in temporarily to hold them while riveting. They are then fastened together by copper nails through sides and headledges, about 1¼ in. apart, the nails being also riveted over burrs. Two or three pieces of wood, as thick as the headledges, are now laid in the trunk to prevent it or the keel from coming together in planking, and are not removed until the boat is finished, or the trunk may close slightly. Now the grooves in the keel are painted with thick white lead, the trunk is driven down into place and clamped fast, rivets are put through the keel and each headledge, then the holes are bored for the screws. These latter are of brass, $\frac{3}{16}$ to ¼ in. diameter and 3½ in. long. The holes are bored full depth with a small bit, then a larger one is run in for a distance equal to the shank of the screw, the latter is screwed firmly in and filed smooth. In fastening such work all joints that are painted must be thoroughly fastened while the paint is fresh, or they will leak. The

CENTERBOARDS.

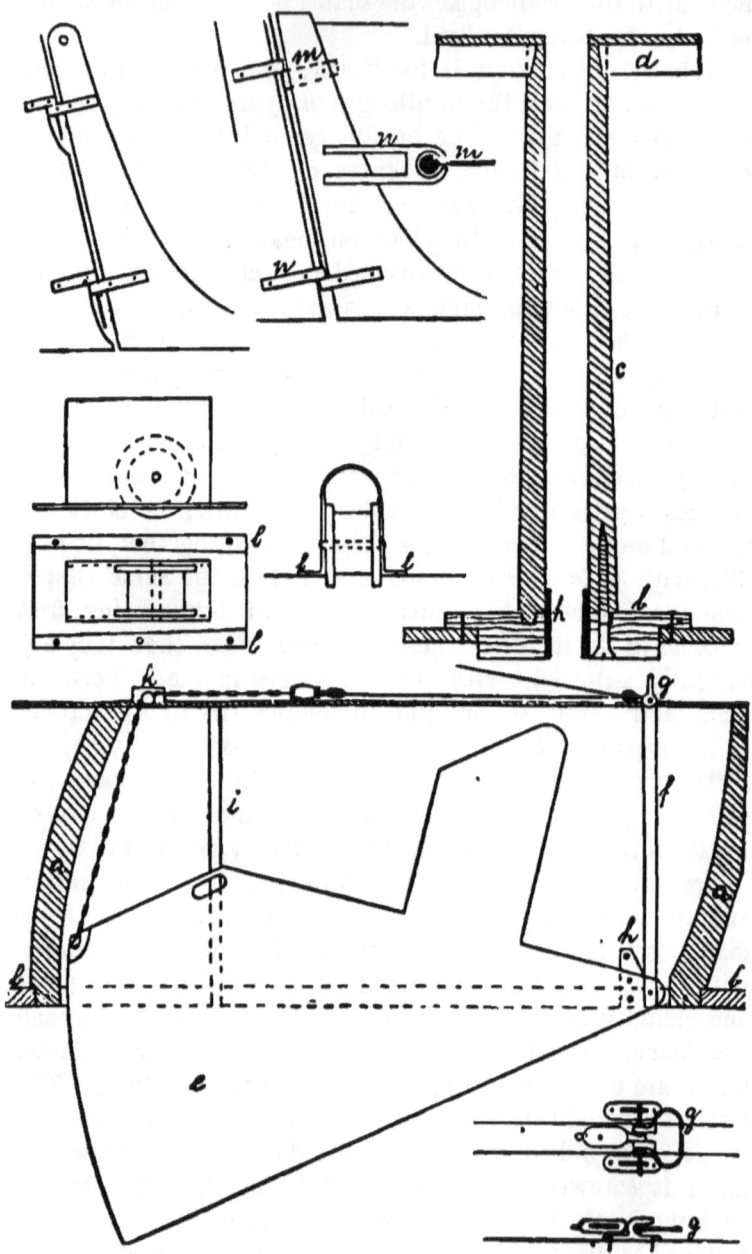

after trunk will come on the deadwoods, and it may be necessary to set in a solid bedpiece, on which the trunk is set, as above, the sides, however, being thinner.

Trunks are sometimes made of galvanized iron, but are liable to rust and are not as good as wood. If the sides of the trunk are thick enough holes are sometimes bored through them from top to bottom and bolts driven down through the keel, preventing them from splitting. In canoes the trunks are usually open on top, so that the boards may be lifted out.

The after board may be of zinc, galvanized iron or copper $\frac{1}{4}$in. thick, about 15 to 18in. long. It is hung by two strips of brass $\frac{3}{8}$x$\frac{1}{8}$in., or even thinner, one on each side of the board, to which they are fastened by a rivet through both and the corner of the board. At the top they are both riveted to a small handle, by which they may be lifted out. A braided cord is used to raise and lower the board, being spliced into a brass eye in the after upper corner. This cord runs over a brass pulley fitted on deck, which is also movable.

The heavy boards are usually of plate iron galvanized, and are from $\frac{1}{8}$ to $\frac{1}{4}$in. thick, the latter weighing 60 pounds. A square board, as is usual in sailboats, would bring too much weight at the top, to avoid which that portion of the board within the case is cut away until only an arm, sufficient to steady the board in the case, is left. The board is first cut to shape out of boiler plate of the required thickness, then it is filed smooth at all corners and angles and reduced to a thickness of $\frac{3}{16}$in. at the upper after corner where the lifting gear is fastened, and at the lower forward corner where the pin hole is. Next the board is galvanized and then it is ready for the fittings. Sometimes cast-iron is used, but it is liable to break. The Pearl canoe has two boards of Muntz metal, one of 68 pounds for racing.

Another form of board, in which the frame work is of wrought iron, with sides of sheet iron, leaving a space inside in which a plate of lead can be inserted, is shown in the Appendix. By this device a light or heavy board can be had, while the weight is divided for carrying. If the board be fixed in the canoe, a brass bolt is put through it and the trunk, on which it turns, but the usual plan is to fit the board to lift out. The board is hung from a brass rod, or between two brass strips (ff), as described for the small board, the top having a handle (g), and also a catch to prevent the lifting rod from pulling forward. A small brass chock (h) is screwed to the inside of the trunk to prevent the lifting rod slipping aft. A rivet is also put through the keel to retain the lower end of the rod. If the board does not weigh over thirty pounds it is raised and lowered by a single pennant of braided cord. Two brass plates are rivetted, one on each side of the board, at its upper after corner, and a brass thimble in which the lifting line is spliced, plays on a rivet through their upper ends. A pulley is placed on deck, from which the cord leads to a cleat.

For a heavier board a purchase must be used, a chain made of flat links side by side, is fastened to the centerboard by two large links, a rubber ball is then slipped on to the chain to act as a buffer, and a single brass block is lashed to the end of the chain. The deck pulley (k) over which the chain runs has a sheave with a square groove to take the chain, and is also fitted so as to slide into place on deck, or be readily removed, without taking it off the chain. A brass block is also lashed to the lifting rod at deck, and the line is rove by making one end fast to the tail of this block, leading through the other block, on the chain, and back through the first block, thence to a cleat. By taking hold of the chain near the pulley with one hand, and of the lifting handle with the other, the pulley may be disengaged and the board readily lifted out.

RUDDERS.

It is most essential to the safety of a canoe that there shall be some means of steering besides the paddle. The boat is so long that it cannot be turned quickly by the latter, the leverage being comparatively short, and on all but the smallest Rob Roys a rudder is a prime necessity. The first canoes were built with stem and stern nearly alike, both with a long curve, to which it was very difficult to fit a rudder. One plan was to use a curved rudder and braces fitted to turn, but such a rudder is not only difficult to ship but will unship itself on the least provocation. In another plan a false stern was made fitting the sternpost, to which it was fastened, but straight on its after edge, to which the rudder was hung. This plan also was clumsy and unsatisfactory, and finally discarded. Another plan was to use a long arm for the lower brace, projecting three or four inches from the sternpost, so that the rudder hung vertically; but this, too, is now little used. For many years the sternposts have been straight, though mostly set at an angle to the keel, as in the old Shadows, giving a good support for a rudder. There was a decided objection to this plan, however, as it was very difficult to launch the canoe from a bank or dock if the water was shoal, the sternpost sticking in the mud and, in addition, it made the canoe hard to turn round. To obviate these objections and yet allow the rudder to hang properly without causing a drag, as it will on a curved or raking sternpost, some canoes of late have had the sternpost vertical, or nearly so, from the water up, giving 7 to 9in. to support the rudder, but below the water the heel is rounded quickly away into the rocker of the keel, allowing the boat to be pushed stern first into mud without sticking fast, and also increasing the ease of turning.

In form the rudder, especially for rough water, should drop below the level of the keel several inches, so as to have a good hold on the water, even when the boat is pitching among waves. With this form of rudder, shown in the large plates of canoes, a tricing line is sometimes used, being made fast to the rudder, and running over a sheave in the sternpost at deck, by means of which the rudder may be raised in shoal water. The idea of a drop rudder in two parts is not new, but its practical application to canoes is of recent date, one of the first having been fitted to the Atalantis by Mr. S. R. Stoddard in 1883. These rudders, now coming into general use, are made of sheet brass, as shown in the drawing, a portion being fitted on a pivot like a centerboard, allowing it to drop to a distance or to rise on striking any obstacle, while it may be raised by a line from the well. This rudder acts, to a certain extent, as an after centerboard, allowing the centerboard proper to be placed further forward than would otherwise be possible. Besides this it has a further advantage, that on most canoes it may be so proportioned as to fold up, leaving nothing below the water-line, thus obviating to a great extent the necessity for removing the rudder at all, as the boat may be launched with the rudder attached, but folded up so as not to strike bottom. If the rudder and yoke are both strongly made, they offer excellent handles by which to lift the after end of the canoe. The stock of the rudder is made of one piece of sheet brass doubled, the rod on which the rudder hangs running down inside the seam as shown. The top of each side is turned down horizontally, and to the two the rudder-yoke is rivetted. The drop portion of the rudder fits between the two sides, a bolt or rivet passing through the three.*

The usual way of hanging the common rudder by pintles and braces, is shown also. A better plan is to have two braces on the rudder, as well as two on the sternpost, with a rod of $\frac{1}{4}$ in. brass running down through them, allowing the rudder to rise up, but not to unship. An old but very good plan is shown at m. On the rudder are two braces, each with a hole through it. A similar brace is placed below on the sternpost and a brass rod is screwed or rivetted permanently into it. The upper end of the rod is held by a flat strip of brass, m, brazed to it, while in the lower brace, n, on rudder, is a slot, allowing it to slide past m on the rod.

The rudder yoke should be strong and well proportioned, as it sometimes receives heavy blows. The arms need not be over $4\frac{1}{2}$ to 5in. long each, as the shorter length will give power enough. Sometimes instead of a yoke a grooved wheel is fitted to the rudder head, the lines running in the groove. This gives control of the rudder in any position,

*See page 198.

even when backing, and has another advantage in that the mizzen sheet cannot foul and the yoke cannot catch in lines or bushes.*

TABERNACLES.

It is now considered necessary in order to spar a canoe to the best advantage, to place the masts so near the ends that it is very difficult, or even impossible to unship them when afloat, especially in rough water. The requirements, both of convenience and safety, however, dictate that they must be capable of being lowered, both for bridges, trees, warps and when in very rough water. The arrangements by which this end is attained are called tabernacles, several styles of which are shown. In one form the deck is not cut, but the heel of the mast is pivoted between two pieces of oak, (p) each $2\frac{1}{2}$x$\frac{1}{2}$in. above deck, fastened securely to the keel and projecting $4\frac{1}{2}$ to 5in. above deck. These pieces are covered above deck with sheet brass $\frac{1}{16}$in. thick, and the heel of the mast is bound with the same to prevent splitting. A pin or bolt of $\frac{3}{8}$in. brass goes through the three, the mast turning on it. The after side of the tabernacle is also of $\frac{1}{2}$in. oak, projecting $1\frac{1}{2}$in. above deck, or enough to catch the heel of the mast and prevent the later from going forward. The mast is raised and supported by a forestay and tackle from the stemhead, to permit which, the sail, if a balance lug, must have a great peak.

Another simple form was fitted to a canoe in 1880 by the writer. A triangular box was set in the forward part of the canoe, fastened at the bottom to the keel, and at the top to the deck, in which a slot was cut, as wide as the mast and about 1ft. long, the box, of course, being of the same width inside. In practice, the canoeist, seated in the well, could place the mast in the box, leaving it, for paddling, lying at an angle of 45 degrees, but when desired to raise it, by going on the knees the mast could be thrown easily into an upright position, and held by a wooden chock (o) slipped into the slot behind it. This chock, with its sides projecting over

*See page 190.

98   TABERNACLES.

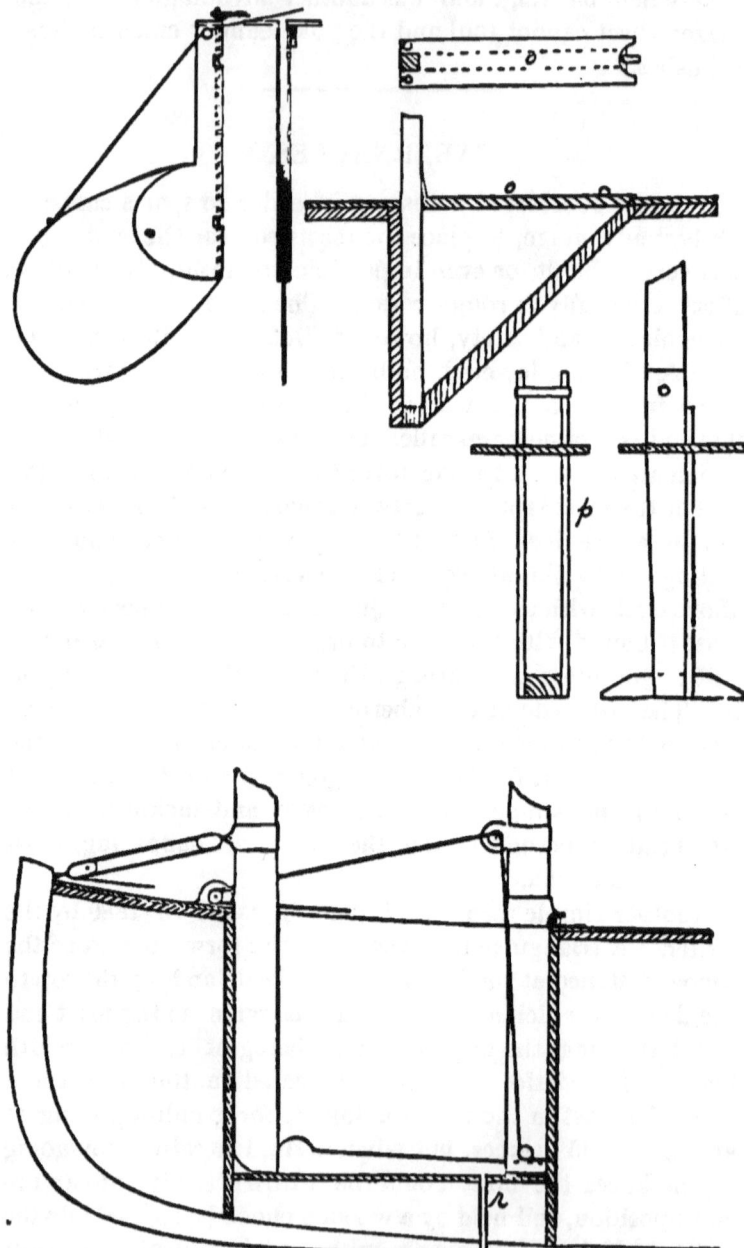

the slot, completely covered it, and kept out all water. When the mast was not in use, its place was taken by a square plug. The chock was fitted at its fore end to slide under two screw heads which held it down, and its after end was kept down with a brass button.

Another and better form of tabernacle is that devised by Mr. Tredwen and fitted to the Pearl canoes. This is a square box 15 to 18in. long, as wide as the diameter of the mast and as deep as can be fitted to the boat. It is lined with sheet copper and provided with a drain (r) at the bottom. For racing purposes two light boxes of wood are made, wide enough to fit in the tabernacle, their united length just filling the remainder of the box when the mast is in it. If the mast is to be set forward, both boxes are slipped in behind it; the mast may be set aft, the boxes being forward of it, or the mast may be placed between the two boxes. By this device the mast may be tried in almost any position until its proper place is found. In the Pearl the large and small mainsails are both used with the same mizzen, the position of the sails being changed so that both will balance properly.

In cruising, the mast is fitted to lower by means of a line from the well. In one method the brass band to which the blocks are fastened is fitted with two lugs or trunnions, at the height of the deck. These lugs engage in two hooks screwed to the deck at the after end of the tabernacle, being raised by a heel rope led over a sheave in the heel of the mast, thence through a sheave on the after side of the mast above deck, and thence through a sheave forward of the tabernacle, giving a very powerful purchase. By another plan the mast is hoisted by a purchase made fast at the stem head and also to the mast above the deck. With either of these arrangements, no forestay is needed. To set the mast at the fore end of the tabernacle, no lugs are required, but a chock is dropped into the bottom to prevent the heel from coming aft, and the purchase is used to bring the mast upright and hold it there.

TENTS AND CAMP BEDS.

A tent of some kind is an essential part of the outfit of every canoeist, as he never knows when it may be needed, even on a short trip. A head wind, foul tide or sudden storm may make it impossible to reach the proposed stopping place and force the canoeist to seek refuge for a night or from the rain as quickly as possible.

Tents for canoes are of three kinds: First, a small shelter, merely for sleeping under; second, a square tent, high enough to sit under and to cook or read in; third, shore tents large enough for two or three. The simplest of all is improvised from a rubber blanket hung over a boom or paddle, one end of which is lashed to the mizzenmast the other resting on the deck. The sides will need to be tied down or kept in place by stones. A better device is the shelter used on the Windward; shown in the drawing. This tent has a ridge rope, one end of which is hooked to an eye or cleat at fore end of well, the other end being made fast to the mizzenmast about 3ft. above deck. The cover is a piece of sheeting or drilling hemmed around the edges, with a hem also down the center in which the ridge rope is run. A triangular piece is fitted to the after end, running across the foot, and tapes are sewn at intervals along the edges to tie down with. This tent, shown with the flap open, makes a good shelter and sheds rain well, but is hardly roomy enough where much sleeping aboard is done. It has, however three advantages, in being easily set and stowed, taking up little room in the canoe, and offering little surface when riding head to wind.

A better tent on the same plan is made with the top triangular, the after end, about 20in. wide, having a hem in which a stick is inserted, a cord from each end of the stick running to the mast. The after end is square instead of triangular, and the sides are triangular, all coming to a point

AMATEUR CANOE BUILDING. 101

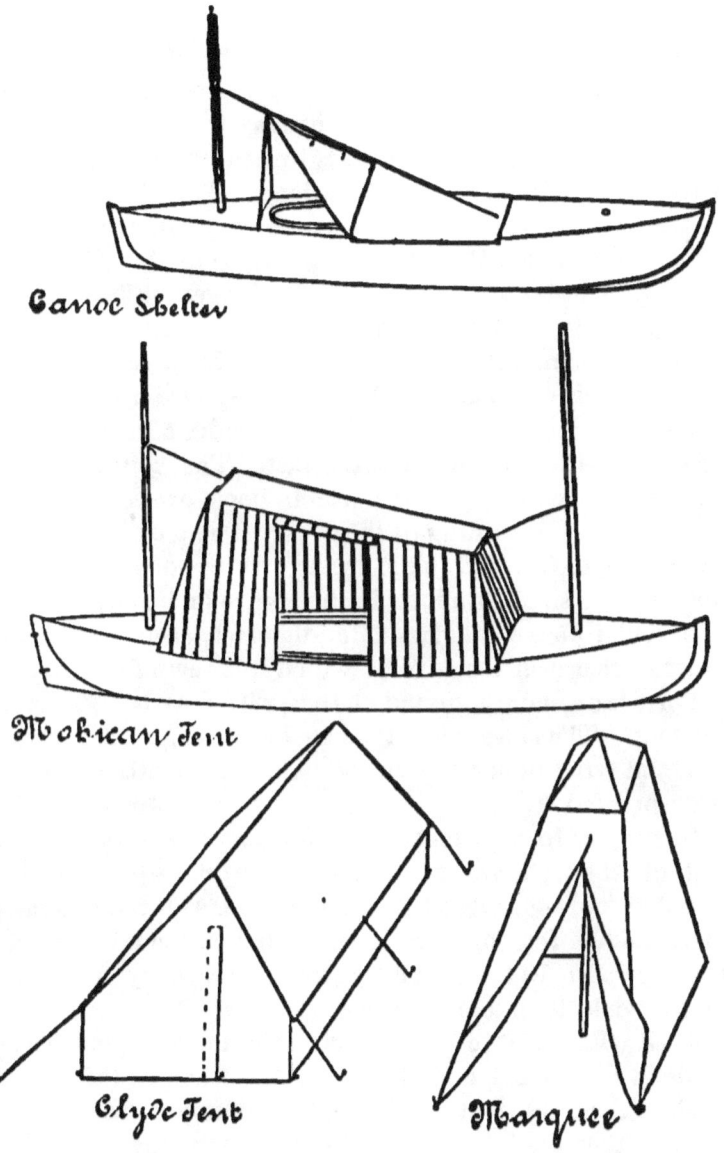

Canoe Shelter

Mohican Tent

Clyde Tent

Marquee

at the fore end. This tent is roomier than the former, but is easily set and stowed.

Of the second class the favorite one is that commonly known as the Mohican, but first used by Mr. C. L. Norton on the Kittiwake. This tent in its present form is also shown. The top piece is of canvas, 22in. wide and 6 to 7ft. long. At each end a hem is turned in, to take a round stick, ¾in. in diameter and 22in. long. The sides and ends of the tent are made of striped awning stuff, which comes 29in. wide, so that three breadths may be used. The tent is 30in. wide at bottom, and about 1ft. longer on bottom than on top. The sides and ends are sewed together at the corners, but the middle breadth on each side is sewed only to the top, making a curtain which may be rolled up, as shown. These curtains lap over the adjoining sides a little, and are provided with tapes to tie them fast. The bottom of the tent is fitted with grommets which hook over small screwheads under the beading of the deck. The tent is supported by two ropes fastened to the masts. It is sometimes desirable to have small windows in the tent, which may be made of circular pieces of glass 2¼in. diameter, each having two holes drilled near the edge by which it is sewn fast.

For use on shore a ground cloth 2¼x7ft. may be used under the tent. The sides should be about 5in. high, to keep out rain and wind under the sides. The floor cloth should be waterproofed.

In another form of tent two bamboo uprights, one at each end of the well, are used, the tent being square, with a rounded top, somewhat like a wagon. A ridgepole, jointed in the middle for stowage, is supported on the uprights, the tent spread over these, and the top extended by four strips of bent oak, let into hems across the top. The Pearl canoe is fitted with a tent of this description, the uprights being made in two pieces, one sliding in the other, so that by extending them the tent is raised, for cooking or reading, but at night they are let down, making the tent lower and less exposed to the wind.

An A tent is sometimes fitted to a canoe, using an upright

at each end of the well, or one at the fore end and the mizzenmast, with the painter stretched across as a ridge rope, but a wider top, as shown in the Mohican tent, is better.

For shore use a tent is usually carried large enough to accommodate two or three persons. The simplest form is the ordinary A tent, made about 6½ft. square at the bottom, and 6ft. high. It is supported by two upright poles and a ridge pole, or the latter may be dispensed with and a ridge rope used, the ends being made fast to stakes in the ground.

A better and roomier form is the wall tent, a very good style being that devised by some of the Clyde C. C. This tent is usually about 6ft. wide, 7ft. long, and 6ft. high, the walls being 2ft. high. The bottom is sewn to the sides and ends, except the flap, which serves as a door, thus preventing all drafts. It is well to have a second bottom of light stuff laid inside over the main one, and not sewn fast, so that it may be lifted out for cleaning the tent. A ridge pole and two upright poles, all jointed, are used. Where the walls join the roof, a hem 2in. wide is sewn, and in this four or five grommets are set to take the tent ropes. The tent pins are of iron rod ¼in., galvanized, 10in. long, with the upper end turned into a ring to draw them out by. A flap is sometimes made in each side of the roof for ventilation. In setting this tent, it is unfolded on the ground and each corner fastened with a pin, then the four pins for the corner ropes are driven, each at the proper distance from its corner, which will be found the first time that the tent is set and marked permanently on one of the poles for future measurement; the corner ropes are made fast to the pins, allowing slack enough to hoist the tent, then the ridge pole is run through, the canoeist goes inside the tent, raises the after end, slips the upright under the ridge pole, walks to the other end, holding up the latter, and slips in the other pole. Now the corner ropes may be looked over and tightened, the remaining pegs driven and the ropes made fast to them, and the ground sheet spread inside. The entire operation, if the tent is properly folded, can be performed by one man in five minutes. Sometimes the ridge pole is made to extend about 18in. beyond the front of the tent, thus keeping the upright out of the way of the door. It is as well to have the rear upright inside, as it is useful to hang clothes on, a few hooks being screwed in it. It will also be convenient to have a few canvas pockets hung to the walls for brush, and comb, etc.

Canoeists in America have used for the past few years a very good tent, of the form known as "Marquee." The ground space may be 7x7ft., the height to peak being about the same. But one pole is needed, which is in the center of the tent. The roof portion may be $2\frac{1}{2}$ft. on each side, and is extended by four small sticks running from the central pole to each corner. The four lower corners are first staked down, the pole is slipped into the center of the roof, raising the latter, then the four sticks are pushed into place, and all is ready. These tents are usually made without a bottom, but a ground cloth should be used in any case.

For small tents, heavy unbleached sheeting may be used, and for the larger ones a light drill or duck. To render them waterproof they may be coated with boiled linseed oil and terebin, one gill of the latter to two quarts of oil, two coats being sufficient. The Mohican tent has a top of heavy canvas and sides of awning stuff, neither being waterproofed, and the marquees are generally made of the latter material. If a stay is made in any place for some time, the shanty tent, described by "Nessmuk" in "Woodcraft," is probably the best known, but in canoeing the halts are usually but for a day or two, and often for a night, so the tent must be quickly set and stowed.

Next to the question of shelter comes the bed, a point of special interest to most canoeists, who for fifty weeks of the year sleep in a comfortable bed at home.

Many canoes are now furnished with a mattress of cork shavings, which makes an excellent bed, and also answers as a life preserver. This mattress, the invention of Mr. C. H. Farnham, is 50in. long, 18in. wide and 4in. thick, made of some light material, such as burlaps or Japanese canvas. It is divided by two partitions, each made of muslin sewn to top, bottom and ends, into three parts, each 50x6x4in., and in each of these about 1¼ pounds of cork shavings is placed. The partitions are intended to keep the cork distributed evenly. Hooks and rings at the ends, with straps for the shoulders, make it easily adjustable as a life preserver, as it is long enough to encircle the body.

In connection with this mattress, Mr. Farnham, much of whose canoeing has been done in cold climates, has devised a sleeping bag or quilt and cover. The quilt, when extended, is nearly heart-shaped, being 7ft. long and 7ft. at the widest part. The small end does not come quite to a point, but an oval end piece is sewn in. The quilt is made of silk or silesia, stuffed with 2¼ pounds of down, evenly quilted in, the edges being strengthened with a binding of tape. Around the edges are buttons and buttonholes, by which the quilt may be converted into a closed bag, in which a man may sleep warmly in the coldest weather. A cover of the same shape is made of fine muslin, coated with boiled oil, and being provided with buttonholes, may be buttoned closely, keeping off entirely the dampness of the ground or even rain. The entire weight of the quilt is 4¼ pounds, and of oiled cover 2 pounds 6 ounces, and both may be rolled into a very small bundle for stowage. The amount of covering may be regulated to suit the weather, the canoeist sleeping with either oiled cover, quilt, or both over him, or if very cold, rolling up in both and lying on the cork mattress. The cork mattress is used in several ways as a cushion during the day. Canoeists usually carry in summer a good pair of blankets, and sometimes a sleeping bag, made of a quilt or blanket doubled and sewn together at the edges and across one end, the other being kept open for ingress.

If weight and space are of importance on short summer cruises, a single good blanket may be taken, with a lining of sheeting or drill sewed to one edge and buttoning along the bottom and other edge. In very warm weather the canoeist sleeps under the drilling only, or if cooler, under the blanket; but in still colder weather the lined blanket will be almost as warm as a double one, and much lighter. A rubber waterbed is sometimes carried and is very comfortable to sleep on, but they are quite expensive.

One or two rubber blankets are usually found in a canoeist's outfit, and are very useful, as a tent may be improvised from one; it is necessary on damp ground or in a wet canoe, and during the day the bedding may be rolled in it. Whatever bedding is carried, it is highly necessary that it should be kept dry, which is best accomplished by wrapping in a waterproof cover or bag, strapping it very tightly, and carrying it well under the deck or in a compartment. In many localities a few yards of mosquito netting are indispensable, as it may be used in connection with any of the tents described. Several varieties of camp cot are sold in the sporting goods stores, but, though good in a permanent camp, they are too heavy and bulky for a canoe.

STOVES AND LAMPS.

On a canoe cruise of any length cooking apparatus of some kind is of course a necessity, but on short trips it is usually dispensed with, a supply of cold provisions being carried. Some means of making tea, coffee or hot soup is always necessary, however, and should be at hand even if the trip in prospect is to last but a few hours. Delays are always possible on the water, and the prudent canoeist will prepare for them. For light cooking an alcohol stove is the cleanest and most compact, the best being that known as the "flamme forcé," which gives a hot flame in a little while, and may be used afloat. With this stove, a little coffee or tea, some pilot bread and a can of prepared soup, a good meal may be quickly prepared. The only objection is the cost of the fuel. Wood spirits may be used instead of alcohol, and is much cheaper; but the odor is very disagreeable. Kerosene stoves have no place on a canoe, as they are so dirty, besides being quite heavy, and the oil is difficult to carry without spilling over the boat. Alcohol for the spirit stove may be carried in a quart can, with a screw top, and even if a little is spilled it will do no injury, as kerosene will.

Most of the cooking will be done on shore over a wood fire, either on the ground or in a camp stove of some kind. Several very compact stoves are made by the dealers in

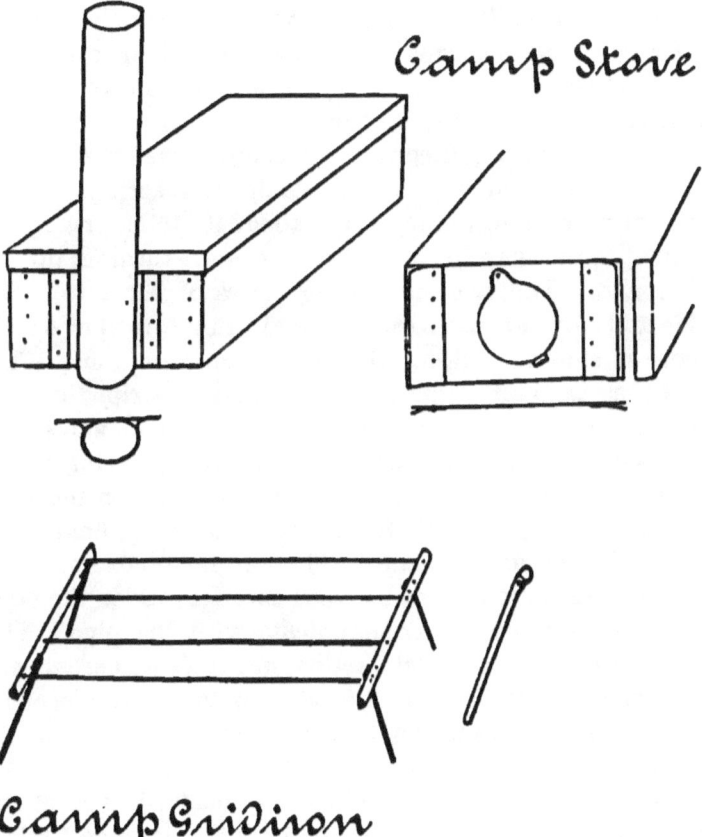

Camp Stove

Camp Gridiron

camp goods, but they are too large for a canoe, unless in a large party, where the load can be divided among several boats. For cooking without a stove a very useful contrivance is the camp gridiron, shown in the cut. The ends are of half round or flat iron 8in. long. Each has four holes drilled in it for the cross bars of $\frac{3}{16}$in. wire, which are riveted in. The legs are of ¼in. round iron, 6in. long, the upper ends being flattened down and turned over to fit on wire staples. These staples pass through holes in the end pieces of the gridiron, and are riveted fast. When in use the fire is made and allowed to burn down to a mass of hot ashes, then the legs of the gridiron are opened and stuck in the ground over it, making a level framework, on which coffee pot, pails and pans will rest without danger of upsetting. When not in use, the legs are folded down and the gridiron stowed in a canvas bag.

A very compact and convenient camp stove was used by Mr. Smith, of Newburg, at the camp last spring. It was made of sheet iron, the top being about 10x15in., or larger if desired, in the shape of a flat pan, the edges turning up 1in all around. The two sides were pieces of sheet iron 6in. wide and 17in. long, 1in. at each end being turned at a right angle, as shown, making the sides each 15in. long. The ends were each 6in. wide and 10in. long, a strip 6in. long and 2in. wide being riveted across each end as shown, on the inside. To put the stove together, the projecting pieces on the sides were pushed in between the strips on the ends, making a square box, and the lid was laid on top, holding all together. In the front end, a circular hole, covered by a door, was made to put in the wood through, and in the other end a hole was cut to communicate with the pipe. This latter is of round or oval section, about 2½in. across, and 18in. long. At the bottom it is riveted to a flat piece 5in. square, which slides in the two extra strips riveted on the after end, as shown in the drawing.

This stove may be easily and cheaply made; it is light and compact for stowage, all folding into a flat package 10x15x-1¼in., except the pipe, and it is quickly set up and taken apart. No bottom is needed, the stove being set on the ground.

In another form the body of stove is hinged together, so that when not in use the stove, covers and funnel all go into a canvas bag, two feet long, one foot wide and about three-quarters of an inch thick, which can be stowed under floor of canoe, and is entirely out of the way. It is made of sheet iron; the top is 24x12, with two holes 8in. diameter, with sheet iron covers, and a small 2x3in. hole at one end to hold chimney or funnel. The sides are 24x10, hinged to top, and ends 12x10, hinged to top in same manner; small strips of heavier iron, ⅛in. thick, are riveted on sides and one end in such manner as to project below bottom of stove, and being pointed, can be pushed into the ground in setting up stove so as to hold all firm. The front end does not have these projections, so it can be propped out from stove, thereby acting both as a door for fuel and to create a draft.

The funnel is made of four pieces hinged together, two 23x3 and two 23½x2½, the additional half inch projecting below and fitting into the hole cut on top of stove.

The top is better in some respects without holes, as the cooking utensils are then kept clean, and free from smoke.

Still another stove is sometimes used, consisting of a cylinder of sheet iron, 10 to 12in. in diameter and the same in length, open at both ends. Across one end are stretched several stiff wires, upon which rest the cooking utensils. At the other end, which is the bottom when used as a stove, an opening about 6x7 from the bottom edge is cut to serve as a door and draft. At the same end, opposite the door, another small opening is cut to give a draft to the other side.

When not used as a stove it is reversed, the wires serving as a bottom enables it to hold all the utensils, plates, etc., as a bucket, and a wire handle being fitted to the bottom for that purpose.

Its advantages are that a fire can be made very quickly, even with poor wood, as the draft is tremendous; it confines the heat and saves fuel, enables one to have a good fire of wood too small to use in an open fire, and renders the hunting and cutting of the usual cross piece for hanging the pots by unnecessary, and it is also very cheap.

To carry the provisions in and keep them dry, a chest of wood or tin is used, generally about 10x15x6in., in which are packed tin cans with large screw covers, such as are used on vaseline cans, for coffee, tea, sugar, flour, oatmeal, baking powder, rice, and any other articles it is desired to keep dry. If the large box be waterproof, as it should be, such articles are sometimes carried in bags of light drilling, but the cans are usually the best. This box is usually stowed just forward of the feet, under the deck, but where it can be easily reached, the spirit lamp being also near by. In cooking on board, the box is drawn out, the lid, or sometimes a hatch, is laid across the coaming for a table, and the spirit stove set up. For cooking on shore, a kettle for boiling water, say two quarts, a smaller one for oatmeal, etc., to pack inside the large one, a coffee pot, and a frying pan are indispensable, other articles being added if there is room. A very handy implement in a camp kitchen is a pair of light blacksmith's tongs, with which plates and pans may be lifted when hot.

A light of some kind is a most important part of a canoe's equipment, as the canoeist may on any trip be overtaken by darkness, in which case his safety may depend largely on his showing a light. A square box lantern of brass is used by many canoeists, one side having a green glass and one a red, the front having a round white lens. The oil used is lard or kerosene. This lamp, which is fitted to slides on the forward deck, makes an excellent signal light, but is not visible from astern. In camp, white slides may be substituted for the colored ones. The use of kerosene is a disadvantage, as it is difficult to carry. The Mohican C. C. carry a small brass lantern in which a candle is used, giving a white light only, and serving for use in the tent or in camp. When under way at night it is hung from the mizzenmast.

CANVAS CANOES.

LONG before the era of boats constructed of boards, and following closely after the primitive attempt at navigation astride a log, and the second step in the form of several logs lashed together, came the intermediate step, by which the form and proportion of a boat was obtained out of comparatively raw material, and without tools. The coracle, as this craft was called, was simply an open frame of basket work, woven from branches and saplings gathered by the riverside, over which the hide of a bullock, or some similar covering, was stretched and sewn; the implements required in constructing such a craft being few and of the simplest form, so that it, in all probability, antedates considerably the canoe fashioned from a hollow log.

This style of boat is still in use, though of course in a greatly improved form, and it still possesses three great advantages, it requires less skill, fewer tools, and less expense of labor and material than any boat of similar excellence. The canvas canoe is usually inferior both in weight, strength and appearance to its wooden rival, but is still a very good boat for all the purposes of the canoeist. The canvas skin is quite heavy when so prepared as to be watertight, and adds nothing to the strength of the boat, which requires, consequently, a stronger frame than a cedar canoe, in which decks and planking add greatly to the strength. If the canoe is of the smaller variety, for paddling only, or carrying but a small sail, it may be built as light or even lighter than a cedar boat of equal stiffness, but if of such a size as 14x30, with 50 to 90ft. of sail, the entire frame must be very strongly braced, and the boat will weigh more than one of cedar.

The first steps of the building are similar to those previously described for a lapstreak canoe. The moulds are cut out in the same manner, the stem and stern are prepared, a rabbet ⅛in. deep being cut to take the edge of the canvas. The inner keel, f, is ⅝in. thick, 2½ to 3in. wide at middle, and tapers to ⅞ at the ends. It is planed up, without a rabbet, and to it the stem a and stern b are screwed. The outer keel is ⅞in. wide, and as deep as may be desired, not less than ⅝in. It is planed up, the grain pointing aft, as described for a cedar canoe, and is fitted to the scarf of stem, and screwed temporarily to stem, keel and stern, as it must be removed when the canvas is put on.

The frame is now set up on the stocks, the moulds shored in place and all adjusted, then the gunwales h, of oak or ash, ¼x⅞, are tacked on and jogs or notches are cut in the stem and stern to receive them, leaving their outer surface flush with the surface of the stem and stern. These notches should not be cut across the rabbets. Strips of oak or ash ll, 1¼x¼in., are now nailed lightly to the moulds, five or six being used on each side, and the jogs dd marked and cut in stem and stern to receive their ends, which, like the gunwales, are secured with screws or rivets to the deadwoods.

The ribs k will be of oak or elm, ⅝x¼in. They are planed up, steamed or soaked in boiling water until quite pliable, and then are taken one by one, bent over the knee, and while still hot the middle nailed down to the keel, and then each ribband in turn, from keel to gunwale, is nailed temporarily to the rib with one nail only. Care is necessary to keep the ribbands fair, without hollows or lumps. After all the ribs are in they must be looked over and faired up, the nails being drawn out, if necessary, after which a copper nail is driven through each rib and ribband where they cross, and riveted, making a very strong and elastic frame.

An inwale, n, 1x¾in , is now put inside of each gunwale, h, being jogged to fit over the heads of the ribs, all three being well riveted together. When this is in, the deck beams o may be fitted. They are cut out of oak or hackmatack, 1x⅝in., and are placed as directed for a wooden canoe, the deck frame and coamings being put in in the same manner. The frame is now taken from the stocks, and all corners that might cut the canvas are smoothed and rounded off, then it is painted all over.

The canvas should be hard and closely woven, wide enough to reach from gunwale to gunwale. The frame is first turned upside down, the outer keel removed, and the middle of the canvas fastened along the keel, with a few tacks, then it is turned over, and the canvas drawn tightly over the gunwales. To do this effectively, the two edges of the canvas are laced together, using a sail needle and strong twine, with stitches about 6in. apart along each edge. This lacing is now tightened until the canvas lies flat over the entire frame. At the ends it must be cut neatly, the edge turned in, and tacked tightly in the rabbet, which is first well painted with thick paint. When the ends are finished the lacing is again tightened up, and a row of tacks driven along the gunwale, after which the lacing is removed and the canvas trimmed down, leaving enough to turn in and tack to the inside of the inwale.

The moulds are now removed, and a keelson, e, is put in to strengthen the bottom, being of oak, $\frac{3}{4}$in. deep and 1in. wide. It is slipped in, one or two of the deck beams being removed, if necessary, and the position of each rib marked, then it is removed, and jogs cut to fit down over the ribs, after which it is replaced and screwed down, running far enough forward on the stem to lap well over the scarfs and strengthen it. The deck frame and coaming is next finished, the mast tubes set, and all preparations for decking made as for a wooden canoe. A deck is sometimes laid of $\frac{1}{4}$in. pine or cedar, over which the canvas is stretched, or the canvas may be laid directly on the beams. The canvas for the deck may be about 6oz. weight, and is stretched tightly down and tacked along the gunwales and around the well. After it is on, half round strips $m\ m$, are screwed around the edge of the deck, and an outside keel piece of oak $\frac{1}{2}$in. thick, is fitted to the bottom, the screws passing through into keelson e, making all very stiff.

The canvas should now be wetted, and painted with two coats of boiled oil, with a little turpentine and japan dryer mixed in, after which a coat or two of paint of any desired color will finish it off. The paint must be renewed on any spots where it may rub off in use, but the canoe should not be painted oftener than necessary, as its weight is much increased thereby.

Another method of building a canvas boat, as described by a writer in *Forest and Stream*, was to build the boat, of whatever model desired, in the same manner as an ordinary carvel built wooden boat, but using very thin planking, no attempt being made to have the seams in the latter watertight. This frame is then covered with canvas laid in thick paint, causing it to adhere to the wood, and making a smooth, watertight surface. Such a boat can be easily built by those who have not the skill and training necessary to build a wooden boat, and it would be strong and durable, as well as cheap.

BOAT BUILDING.

THIS first steps of boat building are the same as those already described under canoe building. The main features of the design are decided on, the drawings or model made, and from them the lines are laid down and the moulds made. The latter, being larger than for a canoe, are usually made of several pieces braced together, as shown in Plate 15, instead of being cut from a solid board. Rabbet and stem moulds will be required, as in a canoe, and also one for the stern or transom, the usual shape of which is also shown, as well as the shape of the mould, which is made of one piece of board, to correspond only to one side of the stern.

One of two methods is usually followed in boat building, either the lapstreak or clincher, as described for canoes, or the carvel or smooth build; the latter being used only where planking is thick enough to caulk, and making a heavier boat than the former. Whichever way is adopted, the boat is usually built on stocks, keel downward; but unless of large size, it will be easier to build it on a table, as described for a canoe.

For a lapstreak boat, a keel or keelson (or if for a centerboard, a flat keel), will be used, as on pages 40 and 42. If the stem is nearly straight, a knee will not be necessary, but the stem may be cut out of oak plank, as at a. The keel c is nailed to it, and the joint is strengthened by a chock e bolted to both. As a boat is usually fuller at the bows than a canoe, the thickness of the stem alone will not give sufficient fastening for the upper planks, so a piece b, called an apron, is added inside the stem, wide enough to fill the space, which the stem alone would not do. This apron is fitted just within the inner rabbet line, and extends from the top of stem down about to the waterline, near which, as the lines become finer, the stem itself will be thick enough for deadwoods. The apron may be from 1½ to 2in. in a fore and aft direction, its width depending on the fullness of the bows.

The sternpost in a boat is of the shape shown at f, the after side being cut away to receive the stern or transom h. The sternpost is nailed or screwed to the keel, and in the angle between the two is fitted the after deadwood g, in which the rabbet is cut. In a lapstreak boat, the keel batten d will run from the chock e, or from the stem, on top of keel and after deadwood to the stern.

The frame being fastened together and the rabbets cut, it is set upon the stocks, the keel is held in place by a few iron nails driven through into the stocks (to be cut off when the boat is removed) and the stem and sternpost are lined up plumb, and with the proper fore and aft rake, and secured by shores from above and below; see page 38, Fig. 9. The transom is next cut out from some hard wood, using the mould b. A vertical line is first drawn down the center of a board of sufficient size, and at its lower end, at 2, the half breadth of the sternpost is set off on each side. A line is drawn at right angles to this center line at the height of the upper side of the gunwales, allowing enough above for the round of the top of the stern, and on this line is laid off the breadth of the stern, giving the points 1, 1. The mould is now applied to one side and then the other, and when both are marked the stern is cut out, allowing enough bevel, as the fore side will, of course, be larger than the after side. The stern is now nailed or screwed to the sternpost, completing the frame.

The moulds are next put in place, and shored from the ceiling or from the floor, and a ridge piece is stretched from stem to stern and nailed to each as well as to the moulds, keeping all in position.

The operation of planking is now proceeded with precisely as in a canoe (see page 45), the stop waters being first put in. The planking should be of cedar, in single lengths if possi ble, but where cedar cannot be obtained, white pine or even spruce may be used. The upper streak is usually of hard wood, oak, walnut or mahogany, and is a little thicker than the lower planking, and is sometimes rabbeted over it, as shown in the sectional view. A bead is sometimes worked near the lower edge, and just above the bead, if a gold stripe is desired, a shallow depression x, called a "cove," is plowed, in which the gold is laid to protect it from injury.

After the planking is completed, the timbers are planed up and put in as in a canoe, or if a neater job is desired, they are made a little heavier at the heels, each one extending only from the keel to gunwale, and are steamed and bent first, then each is fitted to its place, marked and cut to fit down closely to the planks, as shown in the section, after which it is riveted in. Between each pair of timbers a "floor" is fitted, similar to the timber, but extending across the keel as high as the turn of the bilge on either side.

After the timbers are in and nails riveted the next operation is to set the gunwales. These are pieces of ash or oak, $i\,i$, running inside of the upper streak, and covering the heads of the timbers, which are jogged into them as shown in the section of upper streak, gunwale and timbers. The gunwales, sometimes called inwales, may be ¾in. deep, 1¼ wide at center and taper to ½in. at each end. They are planed up, and if necessary steamed until they will bend easily; then they are put in place resting on the heads of the timbers, which latter have been cut off ¼in. below top of upper streak, and the position of each timber is marked. The gunwales are then removed and the jogs cut, after which they are replaced and fastened by a nail through the upper streak at each timber and one or more between the timbers.

After the gunwales are in, a breasthook l, worked from a knee, is put in the bow, fitting the inner sides of the gunwales and the after side of the apron. A rivet of ¼in. iron is put through stem, apron and throat of breasthook. At the after corners, transom knees $k\,k$, are put in, being riveted to the transom and also to the gunwale and upper streak. An oak bead, half round in section, is usually run round the upper edge of the upper streak to complete it, being nailed through into the gunwale.

The interior arrangements of the boat depend on the taste of the builder, but that shown is the usual one in rowboats. In the bows is a small, triangular seat *n*, amidships are one or more thwarts *o o*, according to the size of the boat, and aft are the sternsheets or benches *p*.

All of these rest on two strips *m*, about 2x½in , which are called the risings, and are fastened to the timbers at a proper height to support the seats, which should be about 7in. below the top of gunwale. The seats in bow and stern are also supported by ledges, and the forward ends of the latter are either long enough to rest on the after thwart as shown, or are supported by brackets. The thwarts should be strengthened by knees of wood *j*, well riveted. Sometimes a single knee is used in the center of a thwart, fitted on it and riveted down; and sometimes two are used, one near each edge. The thwart in which the mast is stepped should be very strongly fastened. Lockers are sometimes built under the seats, but their construction is simple, and requires no special direction.

The floor is usually composed of several pieces. in the center the "bottom board," *q*, of about 12in. wide, resting on the ribs and held down by buttons or staples in the keelson; outside of this the button boards *r r*, 3 to 5in. wide at center and narrower at the ends. Several small strips are nailed across the under side of these to keep them from splitting, which strips project ½in. from the inner edge, so as to enter below the bottom board and hold down *r r*. Outside of these pieces are two strips *s s*, about 3in. wide, and screwed to the timbers. They are called the footlines, and on each are two buttons, which turn over the outer edges of the button boards, holding them down. Outside of each footline, and also screwed to the timbers, are the racks *t t*, to hold the stretchers for the feet when rowing. Where the floor narrows up in the stern it is raised a little, one wide piece, *u*, being fitted, resting on two ledges screwed to the bottom.

There are many patterns of rowlock in use, of brass or galvanized iron, and the old wooden thole pins are little used for pleasure boats. The center of the rowlocks should be from 9 to 10in. aft of the edge of the thwart. The rudder will be hung as in a canoe, and fitted with a yoke and lines for rowing and a tiller for sailing. A backboard, v, is usually fitted across the stern, making a back to the seat. The name of the boat may be painted or carved on it. The stem is protected by a stemband of half-round iron or copper, running well down on to the keel, and the angle at the heel of the sternpost is usually protected by a similar piece, called a seagband. The final processes of finishing and painting have all been described in canoe building.

The construction of a carvel built boat varies somewhat from a lapstreak, the operations resembling more those employed n ship building. The frame is prepared as for a lapstreak boat, except that no keel batten is needed. The rabbets being cut and the frame set up, the moulds are put in place and a number of thin ribbands tacked over them. Now, instead of the planking being laid the frame is first set up complete. If the timbers are to be bent, as is usual for small boats up to sailboats of 25ft. or over, a timber block is made of a little greater curvature than the midship mould. The ends are cut from a 10-inch board and cross pieces are nailed to them, making a width of 2 to 3ft. A strip is nailed across each end, projecting a few inches, and to these two ends another piece is nailed, leaving room to insert the heels of the timbers to be bent. The timbers are sawed out and planed up, each being long enough to reach from the keel to the gunwale. They are about one-third deeper at the heel or lower end than at the head; for instance, $\frac{3}{4}$in. deep at heel, and $\frac{1}{2}$in. at head. It is well to get them out and bend them in pairs, that is, if the timber is to be $\frac{3}{8}$in. thick, $\frac{3}{4}$in. deep at heel, and $\frac{1}{2}$in. at head, the piece will be $1\frac{5}{8}$in. wide by $\frac{3}{4}$in. at one end, and $\frac{1}{2}$in. at the other. This piece is steamed and bent on the trap, then sawed in half and the edges planed, making two pieces each $\frac{3}{8}$in. thick.

A steam-box of some kind is necessary for this work, the size depending on the dimensions of the boat. Steam may be made in an iron kettle supported over a wood fire in any convenient manner. A wooden lid is fitted, with a pipe also of wood, leading to the steam chest. This may be made of four pine boards, being 8ft. long and 8x10in. square inside. A light rack of lath is made to slide inside, on which to lay the pieces to be steamed. One end is closed permanently, and the other is fitted with a door, or a bundle of rags is stuffed in, to confine the steam. The timbers being ready, they are laid on the rack and slid into the box, which must be full of hot steam, and left there until they will bend easily. They are then removed one by one and bent over the timber block, the heels first being inserted under the cross-piece, then the heads slowly and carefully bent down, and fastened with a cord, a screw-clamp or a nail. Of course the timbers in various parts of the boat will vary in curvature, but all may be bent on the one block, some being pressed down closer than others. When they are cold they are removed from the block, and before recovering their shape are stay-lathed, a strip called a stay-lath being nailed across to prevent the piece straightening out.

All the timbers are treated thus, and left to cool. Each pair must be marked in some way to prevent confusion. The timbers do not cross the keel, but meet on it, and to join them a floor timber is placed next to each pair. The floors may be sawed from straight stuff in some cases, but toward the ends, and at the middle also if the boat is sharp, they must be cut from grown knees. If the boat has been properly laid down on the mould floor, the floor timbers are taken from the lines on the floor, each being sawed to the proper shape and fastened to the keel by a nail or bolt of round iron (not a screw bolt with nut). After the floors are in place, the timbers are taken, one pair at a time, and fitted in their respective positions. Some will not coincide exactly with the lines of the ribbons, but they may be made to do so by straightening them out a little.

The tendency of bent timbers is to straighten out, so all are bent to a little greater curvature than the ribbands require, and in fitting are allowed to straighten a little. Every timber must touch all the ribbands, or there will be an unfair spot that cannot be remedied, as the timbers are too light to allow any cutting away. The timbers are nailed to the keel and the floor timbers, and also to a few of the ribbands to hold them in place, all being carefully set plumb, and square to the keel.

The widths of the planks are next laid off on the timbers, and stem and stern, no allowance for lap being necessary, of course; and a spiling is taken, not for the garboard, but for the wale or upper streak. This is got out and nailed to the timbers, and the streak below it is also put on; then the boat is taken from the stocks, turned over, and the garboards put on. The planking will be thicker than for lapstreak, not less than $\frac{1}{4}$in., which is as thin as will stand caulking. After the garboards are laid, the broadstreaks follow, then the planking is continued from top and bottom alternately, until an opening is left on the bilge for the last plank, which is called the shutter.

When this is in and fastened, the nails are driven home and riveted, the inside work completed, the bottom roughly planed off, when the seams are ready for caulking.

This operation is performed with a wide, blunt chisel called a caulking iron, and a wooden mallet. The iron is driven into the seam, opening it slightly, then a thread of raw cotton is driven in, using the iron and mallet. On small work, cotton lampwick is used instead of raw cotton. To caulk a boat properly requires care and practice, and the amateur, in default of practical instructions, will do well to employ a caulker. After the seams are caulked they should be well painted over the cotton, using a very narrow brush, as the paint will help to keep in the cotton.

The hull is next planed smooth, sandpapered and painted, after which all seams and nail holes are puttied, all is well sandpapered again, and painted with two coats.

If the boat is to have a deck and waterways, as shown in some of the designs, no gunwale will be necessary; but the upper streak will be heavy enough to take the fastenings at the edge of the deck. A clamp or shelf will be worked in place of a gunwale along the timbers inside, and low enough for the deck beams to rest on it. These beams will be fastened to upper streak and clamp with knees on each beam about the mast. The deck may be of ⅜ or ½in. pine, either painted or covered with canvas. The dimensions of the boat given in the illustration are as follows: Length over all, 14ft.; beam extreme, 4ft.; depth amidships, 17in.; sheer forward, 7½in.; sheer aft, 5in. Waterlines, 3in. apart. The waterlines are drawn for convenience parallel to the keel, but the actual draft of the boat will be 7in. forward and 9½in. aft. Keel outside, 1in.; keel, stem and stern sided, 1¼in.; keel batten, ⅜x2½in.; timbers, ⅜x¾in.; spaced 12in., with bent floors between each pair of timbers; planking, ⅜in.; upperstreak, ½in.; gunwale, 1in. deep, 1¼ wide amidships, ¾in. at ends.

APPENDIX.

COMPARISON OF CANOE ELEMENTS.

AS AN AID to the amateur designer in deciding on the proportions of his craft, the following table has been compiled, giving the elements of some of the best known boats. The first column gives the length on waterline, the second the beam at loadline, and the third the ratio of length to beam or $\frac{L}{B}$. The fourth column gives the distance of the midship section from the fore end of waterline, and the fifth and sixth give the proportions of fore body and after body. The seventh column contains the product of the length on waterline, beam, and depth from waterline to rabbet, giving an approximate comparison of the displacements. In order to compare the relative fullness of the various models on the waterline, the forward part of the waterline is divided in half, and a line K C, Plate I. is drawn and measured. This line is called the "middle ordinate," and is greater as the waterline is fuller. Column seven shows the ratio of this middle ordinate to the extreme breadth of the waterline, the average being about .36. As a further comparison, a dividing buttock line $v\,r$, is drawn in the half breadth plan, parallel to the centerline, and mid way between it and the greatest beam. This line intersects the waterline in two places, r in the forebody, and v in the afterbody. The distance of each intersection from its respective end of the waterline, or the lines $r\,s$, $t\,v$, are measured, and columns eight and nine give respectively the ratios of $r\,s$, to the length of forebody, and $t\,v$ to the length of afterbody. By the aid of these three columns the waterlines may be run in readily in the preliminary drawing. All measurements, for the convenience of calculation, are in feet and decimals.

Elements of Canoes.	Length on loadline.	Beam at W. L.	Ratio of length to beam.	Middle section from fore end.	Ratio of forebody to L. W. Line.	Ratio of afterbody to L. W. Line.	L×B×D.	Ratio of middle ordinate to beam.	Intersection of bow and waterline.	Intersection of buttock and W. L.
Nautilus, Cruising. 1830	14.	2.33	6.	7.	.50	.50	13.9	.36	.39	.30
Nautilus No. 8, Racing, 1879	13.66	2.75	5.	8.	.59	.41	33.8	.36	.34	.26
Nautilus No. 9, Racing and Cruising, 1881	13.9	2.75	5.	7.5	.54	.46	19.1	.36	.34	.25
Nautilus No. 5, Racing and Cruising, 1874	12.8	2.75	4.95	7	.54	.46	18.5	.35	.36	.32
Pearl No. 3, Cruising. 1882	15.	2.63	5.7	7.5	.50	.50	16.6	.38	.33	.33
Pearl No. 5, Racing and Cruising, 1880	14.	2.6	5.4	7	.50	.50	15.3	.35	.29	.29
Pearl No. 6, Racing and Cruising, 1882	14.	2.53	5.3	7.	.50	.50	19.9	.35	.36	.34
Clyde Wren, Cruising. 1879	13.5	2.22	6.1	6.75	.50	.50	12.4	.30	.47	.46
Clyde Laloo, Racing and Cruising, 1881	15.83	2.4	6.6	8.	.50	.50	19.0	.37	.37	.25
Shadow, Cruising, 1878	13.75	2.42	5.7	6.65	.48	.52	14.0	.26	.37	.43
Rob Roy, Cruising. 1867	13.1	2.0	6.5	6.55	.50	.50	9.7	31	.39	.44
Jersey Blue No. 1, Cruising, 1878	13.	2.42	5.4	7.55	.58	.42	13.2	.35	.38	.46
Jersey Blue No. 2, Cruising, 1880	13.45	2.42	5.56	6.55	.48	.52	16.8	.37	39	.40
Raritania, Cruising, 1882	12.72	2.22	5.72	6.32	.49	.51	12.6	.27	.29	37
Kill von Kull, 1880	17.	1.96	8.9	8.5	.50	.50	12.0	.49	.32	.30

MEASUREMENT RULES.

AMERICAN CANOE ASSOCIATION.

RULE 1.—A canoe to compete in any race of the A. C. A. must be sharp at both ends, with no counter stern, or transom, and must be capable of being efficiently paddled by one man. To compete in A. C. A. paddling races, it must come within the limits of one of the numbered classes, I., II., III., IV., and to compete in sailing races, it must come within the limits of either Class A or B.

CLASS I.—*Paddling.*—Any canoe.

CLASS II.—*Paddling.*—Length not over 15ft., beam not under 26in. Depth not under 8in.

CLASS III.—*Paddling.*—Length not over 16ft., beam not under 28in. Depth not under 9in.

CLASS IV.—*Paddling.*—Length not over 16ft., beam not under 30in. Depth as in Class III.

CLASS A.—*Sailing.*—Length not over 16ft., beam not over 28in.

CLASS B.—*Sailing.*—Length not over 17in., with a limit of 28¼in. beam for that length. The beam may be increased ¼in. for each full inch of length decreased.

The greatest depth of a canoe in Classes A and B, at fore end of well, from under side of deck amidships to inner side of garboard next to keel, shall not exceed 16in.

In centerboard canoes the keel outside of the garboard shall not exceed 1¼in. in depth, including a metal keel band of not over ¼in. deep. The total weight of all centerboards shall not exceed 60 pounds; and they must not drop more than 18in. below the garboard; when hauled up they must not project below the keel except as follows: Canoes built before May 1, 1885, may be fitted with centerboards which, when hauled up, may project below the keel, provided such projection of board and case is not more than 2¼in. in depth below the garboard, and not more than 36in. in length. In order to be admitted in races without ballast, the centerboard or boards, including bolts and other movable parts, but not including fixed trunks or cases, must not exceed 15 pounds in total weight.

Canoes without centerboards may carry keels, not over 3in. deep from garboards, and not weighing more than 35 pounds. Leeboards may be carried by canoes not having centerboards.

MEASUREMENT.—The length shall be taken between perpendiculars at the fore side of stem and at the aft side of stern, the beam at the widest part not including beading, which shall not, in Classes A and B, exceed 1½in. in depth, any beading over this depth being included in the beam. The word "beam" shall mean the breadth formed by the fair lines of the boat, and the beam at and near the waterline in the paddling classes shall bear a reasonable proportion to the beam at the gunwale. The Regatta Committee shall have power to disqualify any canoe which, in their opinion, is built with an evident intention to evade the above rules. As the minimum in Class IV. coincides with the maximum in Class B, a margin of ¼in. is to be allowed in measuring for these classes, in order that a canoe built to come well within one class may not thereby be ruled out of another.

ROYAL CANOE CLUB.

Canoes for paddling races must not be of greater length, not of less beam and must be of the material and construction set out in the following classes:

FIRST CLASS.—Any canoe.

SECOND CLASS—(Rob Roy).—Any material or build, decked with wood; greatest length not more than 15ft., greatest beam not more than 26in.

THIRD CLASS—(Rob Roy).—Clinker built, of any material, decked with wood; greatest length not more than 15ft., greatest beam not less than 26in.

Canoes for sailing races shall not be over the following dimensions, viz.:

FIRST CLASS.—Any material and build; greatest length over all, from stem to sternpost, not more than 20ft., with a limit of beam of 2ft., but the beam may be increased by $1\frac{1}{4}$in. for each whole foot of length decreased; greatest depth at fore end of well, under the center of the deck to the garboard, not more than 16in. Fixed keel of wood, not more than 2in. deep; a metal band not exceeding $\frac{1}{2}$in. in depth, may be added to the wooden keel, in which case the depth of the keel inclusive of band must not exceed 2in.

One or more boards are allowed of any material, thickness not exceeding $\frac{4}{8}$in.; length, combined if more than one, not exceeding half the canoe's length; depth of drop not exceeding 18in. below the fixed keel or its metal band. When hauled up they must be completely housed within the canoe.

All ballast, anchors or other metal weights (except centerboard and keelband before described, and metal deck fittings) shall be carried within the canoe, above the garboards. Ballast may be shifted during a race, but all ballast on board at starting must be carried throughout the race.

Second Class Cruising Canoes.—The canoes in this class shall not exceed the dimensions of First Class. Keels and centerboards as in First Class. Weight of canoe, including all spars, gear, fitting and ballast, not over 200 pounds.

STEERING GEAR.

NO detail of the fittings of canoes is as important for safety and comfort, as that by which the rudder is controlled, and no part is so often ill-contrived and badly fitted up.

The strong and simple tiller of the sailboat cannot be used, owing to the distance of the crew from the stern of the boat, and also to the necessity of using the feet for steering, the hands being fully occupied with the sheets, paddle and centerboard. To be of any real use, the footgear must be strong, as a very heavy strain is often thrown on it involuntarily by the powerful toggle-joints of the knees, and the failure of any part, when in rough water or in rapids, might bring disaster to the boat and crew.

The action of the rudder must be prompt and certain without lost motion, there must be a firm bearing for the heels in paddling, and for the ball of the foot in steering, and both must be readily adjustable to suit the length of the leg of the crew. As canoeists know, it is often a great relief, when in the canoe for a long time, to slacken out the footgear, and lean back easily while sailing or paddling slowly, but as soon as a hard paddle is in prospect, the body is settled upright against the backboard, and the footgear shortened up until the feet are braced firmly against it for a long, swinging stroke. The footgear must also be so arranged as to be readily removed for sleeping, stowing luggage, or to carry a second person, and, if possible, it should be so fitted that the second man can steer while paddling.

From the days of the earliest canoeist to the time of McGregor, the paddle only was used for steering, either held in the hands or resting in a small rowlock on either side, called a crutch—Fig. 1—a plan that answered well with the small sail then used; but with the greater number and area of sails something more became necessary, and rudders were fitted, controlled by a continuous line passing along the deck and around the fore end of the well, a pull on either side steering the boat. The increased work thrown on the hands by the addition of ballast, centerboard, spinnaker, etc., made it necessary to transfer the steering to the feet, which had hitherto been idle, so the rudder lines were run through the coaming into the well, and loops tied in the ends into which the feet were inserted, an arrangement still further improved by the addition of metal stirrups. This gave a very powerful and sensitive gear, and it was not in the way in the least, but there were some serious defects in it; there being no brace for the feet in paddling, the stirrups were apt to slip off at times when it was impossible to stop and lean forward and replace them, while in case of an upset the lines might not free themselves and would entangle the canoeist's feet. In one case a canoeist, forgetting to loosen his feet, leaped ashore suddenly and was thrown flat in the water by the rudder lines.

A much better plan was devised by Mr. Baden-Powell for his first Nautilus canoe. As shown in Fig. 3, a vertical spindle of wood has its lower end fitted to turn in a step on the keelson, the upper end running through the deck, the projecting portion being square. Below the deck a crossbar, called the "foot yoke," was fitted to the spindle, and above deck a second crosspiece, the "deck yoke," was fitted to the square head, the rudder lines running along the deck to it. This gear was used for a long time on the Nautilus and all its descendants, and is still often met with. It had many defects, there was no brace in paddling, its position, once fixed, could not be changed, so it was usually just too long or too short for the crew, it was in the way in stowing, sleeping, or carrying double, the lines on deck added to the confusion there, and the parts required careful fitting, and brass bushings at the joints, or they soon worked loose. Several of these objections were removed by some ingenious canoeist, who cast aside the deck yoke, lengthened the foot yoke and ran the lines inside the well, to the extremities of the latter.

The gear shown in Fig. 4 was first fitted to the canoe Janette, in 1877. Two pieces of wood, each 1½x1¼in. and 10in. long, were screwed to the bottom on each side of the keel, running fore and aft, each piece having four vertical notches to receive the stretcher, a piece of oak ½in. thick.

On the foreside of the stretcher a piece of oak 1in. square was screwed, the upper end rounded for the foot yoke to pivot on. The stretcher could be slipped into either of the four pairs of notches, and was then held down by a hook and a screweye in the keel. This gear gave a firm rest in paddling, it was strong in construction, there was no lost motion, and it could be quickly shifted (to make room for a second person) to a pair of similar notches placed forward.

An improvement on this plan is shown in Fig. 5, in which the two fore and aft pieces are grooved on the sides facing each other, and a piece of oak ⅜in. thick and 6in. wide is fitted to slide freely between them. To this piece the stretcher or footpiece is fastened, and in the angle between them is a brass knee or brace, shown separately, the top of which forms a pivot for the footyoke. An eye is cast on the afterside of the brace, in which a short lanyard is spliced. This lanyard reeves through a screweye in the keel, and by it the gear may be held in any position, or by casting it off, the entire piece may be removed. Another pair of slides can be fitted forward or aft, as may be desired for carrying two. This gear seems to fulfill every requisite, and has thus far answered well wherever tried.

The canoe Raven has a novel arrangement, shown in Fig. 6, consisting of two wooden pedals hinged at the bottom to a brass rod, a rudder line being attached to the outer corner of each. A stout brass spring maintains a constant tension on the pedals, and is so formed as to hold them flat on the floor when the rudder lines are cast off. By this arrangement the rudder is always kept amidship when left to itself. The brass rod is held in two holes in the fore and aft cleats, and may be adjusted in the other holes as shown. A better plan would be to hinge the pedals with the spring on a board sliding as in Fig. 5, for which purpose the ordinary spring butts of brass answer very well.

The steering gear, shown in Fig. 7, in which the foot yoke is carried on a spindle passing through and supported by an arched piece of wood, the lower end resting in one of several holes in the floor, was devised by Mr. Rushton. The ends of the arched piece slide in grooved pieces on the floor, and by pulling up the spindle the gear may be slid forward or backward, the spindle end being shipped again in one of the holes.

Where there is a centerboard in the canoe the footyoke is pivoted in a bracket on the after end of the trunk, in which case its position is fixed, and the length can only be changed by using a straight, concave or convex yoke. The Pearl canoe is fitted with a yoke attached to the trunk, Fig. 8, but in order to steer with the feet when lying down, as is done in sailing to windward, the yoke has two loops of leather fastened to its fore side, in which the feet are inserted.

In most of the match-sailing in this country the crew is seated on the deck and the footgear is out of reach. To steer from the deck, a tiller, shown in Fig. 9, is used, having been first applied to the Dot in 1879, and since fitted to many other canoes. A yoke is pivoted on deck just aft of the hatch, and to this yoke a short tiller is fastened within easy reach of the hand. Two short lines join the ends of the deck and rudder yokes. All parts of the gear require to be made very strongly, as a great strain is sometimes thrown on the tiller by the weight of the body. The tiller is sometimes fitted to pivot on the mizzen mast, and is so arranged that a turn of the handle clamps it fast, in any position. Another device for steering by hand was applied to the Folly, S. F. C. C., by her owner; a half yoke only is used on the rudder, Fig. 2., with a stud in the end. A pole long enough to reach the well has a ring in one end, which is slipped over the stud, a push or pull on the pole moves the rudder. A lanyard on the fore end of the pole is belayed to a cleat and keeps it from going adrift if dropped suddenly. This gear is used in the left hand, and is not well adapted to steering from deck.

Another device, only mentioned to warn canoeists against it, has a single stirrup on one side, with a powerful spring on the other. Should the foot be suddenly removed from the stirrup the rudder is drawn quickly to one side and held there. The proper material for rudder lines has long been a subject of dispute among canoeists, and is still undecided, some advocating copper wire, some chain, some a rope of brass or copper wire, and some a braided or hard-laid cord, the last being probably the best, if well stretched and oiled. It will work easily and without the disagreeable clang of wire, and will not kink as chain will do. Whatever material is used, it should lead as directly as possible from the foot-gear to the rudder, with no sharp turns, and holes and screweyes through which it passes should be perfectly smooth. The rudder lines are in some cases run through brass tubes below deck, but this is seldom necessary, and they are best led in around the after side of the well coaming. Some means of taking up the slack in the line is necessary, the usual way being to use a small "fiddle" similar to those used for tent ropes, as in Fig. 4. If obtainable, small snap-hooks should be used to attach the lines to the yokes.

When in use, the steering gear should be examined often, the parts oiled, new lines put in if required, and all parts kept in perfect order. Before a race, of course, it will receive special attention, any parts that appear weak being strengthened for the occasion. Such care is never thrown away, and it is from the lack of just such attention that provoking mishaps occur.

THE FARNHAM APRON.

MR. FARNHAM offers the following additional instructions concerning the apron described on page 63.

1. Carline wires are bent so as to form a hook outside the beading of the coaming, but they do not hook under this beading, for they would then prevent the apron from coming free in case of a capsize.

2. If the forward tube were fastened directly to the coaming, as shown, the apron could not be pushed far enough forward to uncover all the cockpit.

3. The apron is not kept down by the ends of the carlines hooking under the beading, but by the elastic in the hem.

4. The latch or catch to keep the apron stretched must be just forward of the last carline f, and not at i. The apron will not readily come free if fastened at i.

5. The apron is better cut $4\frac{1}{2}$ to 5 inches larger each way than the coaming. The hem is then wide enough to give two thicknesses of cloth along the top of the coaming, where the wear is greatest.

THE WINDWARD CANOE TENT.

A BETTER tent than the one described on page 100 is now used on the Windward. The top is a triangle, the width at the after end being 2ft. The sides are also triangular, about $2\frac{1}{4}$ft. wide at after ends. The fore ends of top and sides meet in a point, which is fastened to the bow of the canoe.

The after part of the tent is square, $2 \times 2\frac{1}{4}$ft., and the upper edge is sewn to the after end of the top, making a hem, in which is a small stick. The tent is hung by a line from the mizzenmast to this stick. The seams, from the ends of the stick to the bow, where the sides join the roof, should be sewn to a light rope, or the sides will be drawn out of shape.

COMPOSITE CENTERBOARDS.

THE following plan for the construction of a centerboard is illustrated in Plate XVIII.:

The sailing canoes of the Royal Canoe Club, of England, frequently carried centerboards of thick iron plate, weighing fifty or sixty pounds. Several canoes, chiefly "Pearls," have recently been built to carry similar heavy centerboards on this side of the Atlantic. A heavy iron centerboard forms most excellent ballast when lowered, but it has some disadvantages. It is unhandy to lift in and out of the canoe, especially if the latter is bobbing about on broken water by a wharf. When fully housed in the centerboard box, it makes a good deal of top-heavy weight, and helps the canoe to roll. Acting on a hint given to me by Mr. W. P. Stephens, at Lake George last August, I have designed, and have had constructed a centerboard loaded with lead, in which the greater part of the weight is concentrated in the lower part of the board. A skeleton frame of bar iron is first made, and on each side of this is riveted a sheet of iron $\frac{1}{16}$ in. thick. This makes a hollow centerboard of a total thickness of $\frac{3}{8}$ in., and weighing 23 pounds. Two light iron frames, $\frac{7}{16}$ in. wide, with long handles, are made to fit into the lower part of the inside of the centerboard. These frames are loaded with lead, and each then weighs 13 pounds, thus making the total weight of the centerboard 49 pounds when fully loaded. This new board will, I think, be found to possess the following advantages:

The weight being concentrated in the bottom of the centerboard gives better ballasting power with less actual weight. I hope that my 49-pound board will give as much sail carrying power as a 65-pound iron plate would.

The lead-loaded board will act as ballast even when housed in the canoe, and will not make the canoe roll, as the greater part of the weight is then within 5 inches of the bottom line of the keel.

For the same reason, most valuable help is afforded in righting the canoe when capsized, even if the board should slip back into the box or had not been lowered. A button across the slot on deck will prevent the board being unshipped in the event of the canoe turning bottom up.

When sailing in shallow water the whole weight of the lead can be got below the keel by lowering the centerboard 8in. only.

It is much easier to handle in removing it from the canoe or putting it on board. Instead of one heavy lift of 50 pounds, you first remove 13 pounds of lead, then lift another 13, and finish with a lift of 23 pounds.

When a light centerboard only is wanted, leave the lead at home. This will in some cases save the necessity of a man keeping two centerboards.

In the accompanying drawing, at figure 1, the board is shown with one side removed, so as to bring the internal economy to view. Starting at F, the frame is continued to E, thence to D and L. It is not continued up to the top corner at C, but is taken across to M, continuing to I, it doubles back in a sort of loop, and is welded at N. This gives doubled strength at M and Bb, where the greatest strain is. The dotted line A B shows the line of the keel when the board is lowered. The portion of the frame from E to about M is made of bar iron $\frac{3}{4}$ deep by $\frac{1}{4}$in. thick. From E to L it is chamfered off to an edge, as shown by the line R R R, so as to cut the water easily. A hook is welded on at P, which hooks on to the king bolt. Where the sheet iron is, the frame is $\frac{1}{8}$in. thick; where not covered with sheet iron—G, I, A, and F to O—it is $\frac{1}{4}$ thick, so as to give a uniform thickness throughout.

The sheet iron is shown by the lightly shaded part. Starting at P, its outer edge passes H, and Aa to G where it is level with the outer edge of the centerboard. It continues past M and C to L, where it takes a jog inward for half an inch to the line R R R. It follows the chamfer along the bottom of the board to near E and up to P again. The upper corner L C M is composed merely of the two thicknesses of sheet iron, and is only an eighth of an inch thick, except where it widens out as it approaches the bar L M. This, while giving plenty of strength, gives room for the chain shackle at C without thinning down and weakening the frame, and it also reduces somewhat the top-heavy weight of frame at the corner. These two sheets of iron, $\frac{1}{16}$in. thick, are riveted to the frame by copper rivets as shown, and are chamfered off from L to D and D to E, to correspond with the chamfer of the frame. From C to L the edges of the sheets are brought together so as to continue the sharp edge. Figure 2 gives a full-sized section at one of the rivets. The two frames which contain the lead are made of $\frac{1}{2}$x$\frac{3}{16}$in. iron, hammered on the edge down to $\frac{7}{16}$, which increases the thickness slightly. They are shown at S, T, U, V, and X, Y, Z, O. The lead is held by pieces of stout wire which are riveted in the frames before the melted lead is poured into them. These wires are shown by dotted lines. The frames are fitted with long handles V I and O J, which terminate in eyes I and J. Above the eye J is a button K, working on a bolt F, secured by two jammed nuts below. This makes it impossible for the lead to fall out if the centerboard is upside down. The eyes I and J project above the deck.

To take out the lead when the board is housed in its box, turn the button K, put your finger in the eye J, and lift out the forward lead frame, then by means of the eye I, drag the aft frame forward, and lift it out.

ROBERT TYSON, Toronto Canoe Club.

The smaller sketch is a suggestion of Mr. King's, in connection with the same design.

DESCRIPTION OF PLATES.

PLATE I.—CRUISING CANOE "JERSEY BLUE."

The first canoe of this name was designed and built by Mr. W. P. Stephens in the winter of 1877-78, being intended for a cruising boat. The dimensions were nearly the same as the present boat, but the model was quite different, with greater sheer, long bow and full quarters. She was fitted with deck hatches, sliding hatch to well, rudder, and the footgear shown in Fig. 4, Plate XVII., and was rigged as a schooner, two boom and gaff sails and jib. The rig was subsequently changed to leg of mutton, and later to balance lug. The model shown in Plate I. and II. was designed in 1880 for the same purpose as the preceding one, general cruising, and a number of canoes have been built from it.

Plate I. shows the lines of the boat, and also method of putting them on paper as explained in the chapter on designing. The dimensions and table of offsets are given on pages 13 and 22. In cruising this canoe will carry a mainsail of 45ft., with mizzen of 18 to 20, and in racing, a mainsail of 65 to 70 sq. ft.

PLATE II.—"JERSEY BLUE," CONSTRUCTION DRAWING.

This drawing shows the arrangement of decks, bulkheads, etc , and the general construction of the same canoe, and is described on pages 52-55.

PLATE III.—RIVER CANOE, "RARITANIA."

This canoe was designed by Mr. W. P. Stephens in 1882, for work on small rivers and streams. She is built with a flat keel, and can be fitted with a centerboard or a false keel of wood can be screwed on. The floor is flat, the keel projects but ¼in., and on each side are oak bilge keels ¾in. square. On to these and the main keel the boat rests squarely, and may be dragged without injury. Two sails are used, either leg of mutton or lateen, the latter being the better. Their areas may be 15 and 30ft. for cruising. Length 14ft., beam 27in., depth amidships 9½in., sheer at bow 3½in., sheer at stern 2½in., crown of deck 3in.

TABLE OF OFFSETS, CANOE "RARITANIA."

	Stations.	0	1	2	3	4	5	6	7	X	9	10	11	12	13	14
Heights.	Gunwale	13	12¼	11⅝	11⅛	10½	10	9½	9½	9½	9½	9¾	10 1/16	10½	11 3/16	12
	Rabbet	13	5⅝	3¼	⅞	....	....	....	....	....	....	....	⅜	1¾	4 11/16	12
	Keel	....	4⅜	1⅝	½	....	....	....	....	....	..	....	⅛	1⅛	3 11/16	.. .
Half Breadths.	Deck	7-16	4	7⅜	10	11¾	12¾	13¼	13½	13½	13	12⅝	11	8¼	4½	7-16
	8in. Water Line	....	1¾	5½	8½	11	12½	13¼	13½	13½	13¼	12¼	10	6⅛	2½	....
	6in. Water Line	....	¾	4¼	7½	10	12	13	13¼	13¼	13	11½	8¾	5	1¼	....
	4in. Water Line	....	...	2¼	5¾	8½	11	12¼	12¾	12¾	12	10⅛	7⅛	3¼	....	..
	2in. Water Line	....	...	...	3⅜	6	9	10⅝	11	11	10	7¾	4	¾	.... ..	....
	Diagonal, A. B.	....	3½	7	9⅝	11⅝	13½	14½	14¾	15	14¼	12⅞	10¾	7¾	4⅜	....
	Diagonal, C. D.	....	⅝	3⅛	5⅜	6¾	7⅞	8¼	8½	8½	8¼	7⅜	5⅞	3⅝	1	....

PLATE IV.—THE SHADOW CANOE "DOT."
This model was designed by ex-Com. W. L. Alden, N. Y. C. C., in 1878, and was built by Mr. James Everson. The Dot was the third of the model and was built in 1878, since which she has been widely known as a successful racing and cruising boat. Her first cruise was from New York to Rondout, in 1878, and in 1880 she made a cruise on the Susquehanna, from Binghamton to Harrisburg, in nine and a half days, since which she has made many short cruises, besides several of some length. Her first race was in the regatta of 1879, in which she was beaten by boats with larger keels. In 1880 the keel was increased to 2¼in., and in 1883 to 3in., which depth is sufficient to take her to windward, as she has won nearly every sailing race in which she has entered, including five for the Challenge Cup, besides winning all of the sailing prizes but one in her class at Lake George in 1882. Her best run on a cruise was fifty miles in ten hours under sail and paddle, from New York down the Sound. Her owner, Mr. Vaux, was one of the first in this country to use lug sails, having two standing lugs, which were changed in 1881 for balance lugs. She was also the first boat steered with a tiller, the crew sitting upon deck. The following are her principal dimensions:

	Ft.	In.
Length over all	14	4
Beam at Waterline		30
Beam at deck		28
Depth at bow		10½
Depth at stern		10½
Depth amidships		1½
Depth of keel		2½
Distance from fore side of stem—		
To forward hatch	1 / 2	6 / 8
To center of mainmast	2	6
To forward bulkhead	3	6
To fore end of coaming	4	0½
To sliding bulkhead	8	11
To after end of well	10	3
To bulkhead	10	10
To center of mizzenmast	11	4
To after hatch	11 / 12	0½ / 8½

Weight of hull when in use, 93 pounds.

TABLE OF OFFSETS FOR CANOE "DOT."

Stations.	Depth at Gunwale.	Half Breadths.				Diagonals.	
		Deck.	L. W. L.	No. 2.	No. 3.	A. B.	C. D.
0	16½						
1	12¾	6¾	7¼	4¾	3⅛	7½	6¾
2	10½	11¼	10¼	9⅞	8	12⅞	10¾
3	9½	13⅛	14	13	11	15¼	12¼
4	9½	14	15	13⅝	11½	15⅝	12¼
5	9¾	13⅞	14½	13⅜	10⅝	15¼	11½
6	11	12½	12¼	10¾	7½	13½	10½
7	13⅛	7¾	6⅛	4¾	3	8½	7½
8	16½						

The keel, stem and stern are 1in. thick; planking (lapstreak, 5 planks on each side), ¼in,; decks and hatches, ¼in.; ribs of oak, ¼x⅝in., spaced 6in. apart. Many changes have been made in the boat as experience has shown them to be necessary; the fore bulkhead, shown by dotted lines, has been removed, the fore hatch permanently fastened down, 2in. of keel added, foremast tube shifted forward and enlarged from 1⅝ to 2in., the old steering gear, with a yoke on deck and one below, replaced by a yoke below deck on a vertical pivot, and the elliptical well entirely covered with hatches changed to one with a pointed, flaring coaming, with an apron. The paddle used for several seasons past has been 9ft. long.

PLATE V.—RACING SAIL OF THE "DOT."

The racing rig of the Dot consists of two balance lugs, of 70ft. and 25ft., the larger of which is shown in Plate V.

DIMENSIONS OF SAILS.

	Main.	Mizzen.
Luff	6ft. 10in.	4ft.
Leach	10ft. 9in.	6ft. 4in.
Foot	9ft. 8in.	5ft. 9in.
Head	7ft. 8in.	4ft. 4in.
Tack to peak	13ft.	7ft. 8in.
Clew to throat	10ft. 7in.	6ft. 3in.
Area	70 sq ft.	25 sq. ft.

Battens 24in. apart on leach and 22in. on luff.

When the sail is taut the ring on the yard is drawn close in to the mast, raising the yard and throwing the fore end a little further forward than it is shown. The halliard *a a*, is hooked into an eye on the parrel, *c*, (the latter made fast to the yard just forward of the mast) from which it leads through a ring on the yard, thence through a block *d*, at the mast head, and down through a ring lashed to the mast, near the deck, from which it leads to a cleat abreast the well. The tack *b b* is seized to the boom just forward of the mast, and leads through a hook on the boom abaft the mast, under a hook in the deck, and to its cleats.

The parrels *e e*, are made fast to the battens just forward and aft of the mast, and when in place, hold the sail in to the mast, keeping it flatter, and relieving the masthead of considerable strain. The reefing gear is rigged as follows: Three deadeyes, *f f f*, are seized to the boom as shown. The reef line *h*, from the leach, is in two parts from the batten to the deadeye, one part on each side of the sail. At the deadeye, they unite into one part, leading forward along the boom, through the middle deadeye, thence through the block *i*, on fore reef line. This line *g* also runs down each side of the sail, through the deadeye, and is then lashed to the single block *i*. A pull on the hauling part (the halliard being first slacked away) brings boom and batten snugly together, the line is belayed to the cleat on the boom, and the middle reef-points *l* hooked together, or a third line may be added in place of the points. A similar arrangement may be rigged on the batten, drawing down a second reef. The points on the halliard where it is belayed when a reef is hauled down are marked with colored thread, so the halliard can be slacked away the proper distance, made fast, and the reef hauled in and belayed. A sling about 18in. long has both ends seized to the boom. On this a deadeye travels, to which the sheet is fastened.

PLATE VI.—CLYDE CANOE "LALOO."

The following description of the Laloo, with the drawings, was furnished by Mr. C. G. Y. King, of the Clyde C.

C., a well-known canoeist, as well as an amateur designer and builder. The design differs in many respects from American models, and has never been tried in competition with them. It will be noticed that the lines, which show the inside of planking, are narrowed in amidships to allow the boat to spread in building.

Mr. King says: Talking one evening over a quiet pipe with an old canoeing friend, Charlie Livingstone, of Liverpool, we both agreed that a new design of canoe was necessary (to our ideas), and if not actually promoting canoeing, it would give us some new experience in canoes. So we set to work sketching free-hand designs, and in course of time hit upon the idea of a canoe having very full lines aft, carrying the floor well forward, so as to give the basis of a full bow which at the same time would look as if it were extra fine. Our aim was to build a canoe that would, for her size, be the stiffest under sail, quickest under paddle, and a good dry seaboat. We succeeded. The lines of accompanying drawing are the inside or skin lines.

To those who do not understand what that means, a few words will explain. The principal dimensions of the canoe are: Length, 16ft.; beam, 31½in.; depth from inside of girboards to top of top-streak amidships, 11¾in.; depth of keel, including metal band, 2¼in. In setting up the frames it is a wise thing to cut them at most 1in. less beam amidship than beam required when finished, as the thickness of the planks each side has to be allowed for, and the boat is almost dead certain to fall out after the tie beams are removed previous to screwing down the deck.

The drawings give a sheer plan, a body plan and a deck plan. The lines A and B in all three are buttock lines. The waterlines are indicated in the body and deck plans by 1,, 2,, 3,, 4,, and cross sections in body and deck plans by 1, 2, 3, 5, 6, 7. The midship section is somewhat different from what is or what the writer knows as the Shadow model. The Shadow has too much tumble home and loses stability as she lies over to a breeze.

The Laloo has her greatest beam at the gunwale, and has

no tumble from bow to stern, thereby increasing her stability from her waterline to her deck, and enabling her to carry an extra amount of sail. Her sail power by calculation is 53 sq. ft., and she is able to carry that spread without ballast. She can carry safely for racing purposes 114 pounds of lead, and with that amount she can carry 85 sq. ft. sail. The best style of sail to have, especially in Scotch waters, is the batten sail with a running reefing gear, which enables the canoeist to reef his sail close down while under way, and without more exertion than hauling on a cord specially arranged for the purpose.

The Laloo's rig is one lug sail of 65 sq. ft., which is a handy size for cruising or racing, and 70 pounds of lead, 40 pounds placed at fore end of well in front of the foot-steering gear, and 30 pounds placed aft the sliding bulkhead at aft end of well. To those who might contemplate building such a craft a few over-all dimensions of deck fittings might come in handy. Length over all, from bow to sternpost, 16ft.; from bow to center of mast step, 2ft. 10in.; from center of mast step to fore end of well, 4ft.; from fore end of well to aft end of well, 3ft. 5¼in. from aft end of well to aft end of hatch, 1ft. 6¾in.; from aft end of hatch to sternpost, 4ft.; width of well at fore end, 1ft.; width of well at aft end, 1.ft.; width of locker hatch at fore end, 1ft. 8in.; width of locker hatch at aft end, 1ft. 1in.; height of well coamings, 1¼in.; diameter of mast at deck, 2¼in.; diameter of mast at head, 1¼in., height of mast from deck, 10ft. To any one studying these lines and comparing them with those of other craft, the difference will be very marked. It was predicted by those who saw the canoe under construction that she would have a heavy drag aft, but such is not the case. She enters the water with perfect sweetness and leaves it without a ripple even when running before a good breeze in a calm sea. Her stability and sail-carrying powers leave no loophole for adverse criticism. To Mr. Livingstone is all the credit due for insisting on carrying out and building these strange lines for a canoe to have. She is easy to paddle considering her 31¼in. beam, and her stowage capacity is most ample for a long cruise. She is perfectly open down below, fore and aft; has no water-tight bulkheads, but has instead probably india rubber air bags fitted to her shape (before deck is screwed down at bow and stern). These bags will be about a couple of feet long, and can be inflated at will, and have more than enough buoyancy to float the canoe when full of water, and with her crew on board.

At the aft end of well is a sliding bulkhead, and by removing it and folding back the hatch-lid H, room can be made for a crew of two; or when cruising alone without a tent a comfortable couch can be obtained in a few seconds.

PLATE VII.—SAIL PLAN OF "LALOO."

This sail has an area of 60 sq. ft.; the first reef has 16 sq. ft.; second reef, 15 sq. ft.; leaving 29 sq. ft. for a close reef when blowing hard. Figure 1 is a full sail plan, showing all the rigging necessary without being complicated. A are main halliard blocks at masthead and foot of mast; D is double block for topping lift; K is single small block for jackstay; J S, jackstay; M H, main halliard; T L, topping lifts—one on each side of sail; B, reefing battens; R, reef points; R C, forward reef cord; R C 2, after reef cord; P, loop and toggle to secure lower end of topping lift; C, cleat to receive R C 2 when reef is hauled down.

S — parrel on boom, on which runs a deadeye or block, to which is fastened main sheet. When lying close hauled the block is at the after end of S, and S helps to distribute strain on boom; when running free block is at fore end of S and main sheet does not drag in the water. M = mast. W on boom and on each batten are parrels to keep sail close up to mast so that it won't bag with the wind. W O, jack block. Figure 2 illustrates on a large scale how to fasten halliard to yard so as to dispense with the services of a traveler. T at the throat is a loop fastened to the yard through which passes a toggle on the end of the halliard. The halliard then passes round the opposite side of the mast, from which yard and sail are, is reeved through block B on yard, then through block A at masthead, then down to block at mast foot and thence to cleat. The topping lifts are toggled to boom so as to be easily detached when spinnaker is to be hoisted, spinnaker head lying ready to be fastened to either as required, the other topping lift remaining in its place.

The jackstay is rigged on the outside of the sail, so that when sail is lowered the triangular part at boom, Y Z, prevents the sail from flopping over the deck on the one side, while the mast prevents it on the other. X is a brass rod at the masthead for a fly. There are several plans by which the sail can be reefed "instantaneously." The one here shown the writer has found to work the best. Let us start at the back and follow the first reef all round. One end of the cord is secured at the back, and is rove through brass rings $\frac{3}{16}$ diameter sewn on sail where shown, then through block at luff of first batten, then through block in line of mast, then down to a deadeye at mast foot, then to cleat wherever handiest for owner. Then the after part has to be looked to. Rig cord in the same way, starting at the clew and cleat on C at boom. This gives a very handy plan for reefing quickly if caught in a squall while racing. When the squall has passed slack out reef cords and hoist yard at once. For a good, deliberate reef while cruising it would be well to tie down reef points as well, as the extra time it takes is not wasted. It is a capital plan to have all blocks for use about the mast fastened to the mast and not to the deck, so that when one comes ashore to dismantle, the mast, sail and rigging can be removed and returned without the bother of always re-rigging.

The Laloo was designed to be sailed without a mizzen, though an after sail is of great service in mostly all weather.

PLATE VIII.—NAUTILUS RACING AND CRUISING CANOE.

For the drawings of this canoe, as well as the two following ones, and the canoe yawls, we are indebted to "Yacht and Boat Sailing." This canoe was designed by Mr. Baden-Powell, for open water cruising and for racing under the R.C.C. rules. The main objects in view were sleeping room, good sailing lines and light draft. Centerboard of plate-iron,

83 lbs. Length, 14ft.; beam, 33in.; depth amidship, 14½in.; sheer at bow, 7¾in.; do. at stern, 5¼in.; draft, 7in ; keel, 1in.

AA—Mast tubes. BB—Headledges. C—Centerboard. D—Fore bulkhead, with door. E—Drain pipes to compartment. F—Footyoke. G—Deckyoke. H—Handle of centerboard. I—Hauling up gear of centerboard. J—Rack for cleats. K—Fore hatch. L—After hatch. M—Seat for paddling. N—After bulkhead, with door. O—Floorboards. P—Backboard for paddling. S—Sheer for rudder tricingline.

	Sections..............	0	1	2	3	4	5	6	7	8	9	10	11	12	13	14
Depths.	Deck to Waterline....	16¼	14½	13	11½	10¼	9½	9	8¾	8½	8½	9	9¾	10¾	12	13¾
	Waterline to rabbet....	....	5	5¾	6	6	6	6	6	6	6	6	6	5½	4½	....
Half Breadths.	Deck..........	....	6	10	12¾	14½	15½	16¼	16¼	16½	16½	16	14¾	12¾	8½	½
	Waterline........	....	3¼	6½	9¾	12½	11¾	15¾	16¼	16¼	16	15	13¼	9¼	4½	....
	3in. Waterline	....	1¾	4¼	7	10	12¼	14¼	15½	15½	14½	12¼	9	4½	1½	....
	Diagonal......	....	4¾	8¼	11¼	13¾	15¾	16¾	17¼	17¼	17	16¼	14½	13½	7	....

The sails are two balance lugs of 95 and 25sq. ft.

PLATE IX.—PEARL CRUISING CANOE.

The family of Pearls, designed by Mr. E. B. Tredwen, R.C.C., numbers nine different models, the design in the plate being No. 3. She is designed for open water cruising as well as racing. Dimensions: Length, 15ft.; beam, 31½in.; depth amidships 11in; sheer at bow, 5in.; sheer at stern, 3in.

	Stations	½	1½	2½	3½	4½	5½	6½	7½	8½	9½	10½	11½	12½	13½	14½
	Rabbet Line to Gunwale.	10	14	13	12¼	11¾	11½	11¼	11	11	11⅛	11¼	11½	12¼	13	14
	Rise of Rabbet Line.	2¼	1¼	½	....	....	...	..	...	..	....	...	...	½	1¼	2¼
Half Breadths.	Gunwale	3½	8¼	12	14	15	15½	15½	15½	15½	15¼	14½	13½	11¼	8	3¼
	9in. W. L.	2¼	6	10	12¾	14½	15¼	15½	15½	15½	15	14	12¼	9½	6	2¼
	6in. W. L	1⅛	3½	7¾	11	13¼	14½	15	15¼	15	14¼	18	10¾	7¾	3½	1⅛
	3in. W. L.	¾	2½	5¼	8½	11½	13½	14	14¼	14	13¼	11¼	8¾	5¼	2½	¾

148 DESCRIPTION OF PLATES.

PLATE X.—PEARL CANOE No. 6.

This canoe was designed to compete not only with canoes, but in the races of the Thames gigs, boats much larger than canoes, and she has been remarkably successful both with them and her own class. She is fitted with two centerboards of Muntz metal, the forward one of 68 pounds, being ⅜in. thick. The sail carried is 105ft. in mainsail, and about 40ft. in mizzen, the latter being fitted to reef, by rolling on the boom. Length 14ft., beam 33in., depth amidships, 14in., sheer at bow 7in., at stern 4in.

Stations	Heights		Half Breadths				
	Rabbet Line to Deck	Rise of Rabbet Line	Gunwale	12in. Water Line	9in. Water Line	6in. Water Line	3in. Water Line
1	17	1					
2	15¾	½	0¼	1¾	3¾	5¼(?)	...
3	14¾	¼	13¾	11¾	7⅞	5¼	3
4	14¼		15	13¾	11⅜	9¼	5⅞
5	14		15¾	14⅞	13⅞	12¾	11
6	14		16	15¾	14⅞	13⅞	6⅞
7	14		16	15⅞	15¼	14⅞	12⅞
8	14		16	15⅞	15⅞	15¼	12⅞
9	14		15¾	15⅜	14⅞	14⅜	12
10	14⅜		15	13⅞	13⅞	12⅞	9⅜
11	14¾	¼	13⅜	12⅞	11⅜	6⅜	0⅜
12	14⅞	½	11⅜	9¾	8¼	0½	3¾
13	15⅜	1	9¼	6¾	5⅞	4⅞	1⅜

Plate XI.—American Cruising Canoe.

This canoe was designed in 1883 by Mr. W. P. Stephens, of the New York C. C., for general cruising and racing.

Stations	Depths		Half-Breadths			
	Deck.	Rabbet.	Deck.	6in.	4in.	2in.
	Ft. In.	Ft. In.	Ft. In.	Ft. In.	Ft. In.	Ft. In.
0	1 0	....	0¹	6¹	....	....
1	1 4⁴	1³	2⁷	2³	1⁶	0⁶
2	1 2⁷	0⁴	7⁴	5¹	4	2³
3	1 1⁴	0¹	10²	8¹	6⁶	4⁴
4	1 0³	0	1 0²	10⁷	9⁴	6⁶
5	11⁵	....	1 1⁶	1 1	11⁶	9²
6	11¹	....	1 2³	1 2²	1 1⁵	11³
7	11	....	1 2⁶	1 2⁷	1 2³	1 0²
8	11¹	....	1 2⁴	1 2⁶	1 2	11⁷
9	11³	....	1 2¹	1 1⁶	1 0⁶	10²
10	1 0	....	1 1	11⁶	10¹	7⁶
11	1 0⁶	0¹	10⁶	9	7⁴	5¹
12	1 1⁵	0⁴	7⁶	5⁶	4⁶	2⁶
13	1 2⁵	1¹	0⁴	2⁶	2	1
14	1 4	....	0¹	0¹	....	....

To make the same lines answer for a 15×30 canoe, the moulds, six in number, may be spaced 25⁷in. apart instead of 24in., as shown. An extra mould at each end, Nos. 1 and 13, will be useful in building.

The movable bulkhead is placed 8ft. 3in. from the bow, and shapes aft, giving an easier position to the body than when vertical. The after bulkhead is placed 9ft. 9in. from the bow, and is fitted so as to be watertight up to the top of the coaming, which runs aft 15in. further, the bulkhead projecting $\frac{3}{16}$in., or the thickness of the hatches, above the coaming. On the top of this bulkhead is screwed a strip of flat brass d, $\frac{3}{32}$in. thick (see Plate XVIII.) and wide enough to project ⅜in. on each side of the latter; thus, if the bulkhead is ½in. thick, the brass should be 1¼in. The cuddy hatch b is $\frac{3}{16}$in. thick, flat, with no crown, and extends from the fore side of the bulkhead to the shifting bulkhead, and may project a little over the latter. In width it extends ⅜in. over the coaming on each side, allowing side pieces ½in. thick to be nailed to it, the latter extending down to the deck. The grain should run athwartship, and the hatch may be strengthened by a batten screwed to the under side, running fore and aft. The after hatch *a* is made in a similar manner, but extends aft of the well ⅜in., with a piece across the end as well as on the sides. The side pieces of both hatches meet at the joint shown.

*Both hatches turn on flat brass hinges with brass pins, which are riveted to the brass strip, and the hatches may be fastened with hooks and screweyes on deck, or with hasps and padlocks. The cuddy hatch is opened by turning it aft, while the after one turns forward, each when open lying flat on top of the other. The side pieces, reaching to the deck, keep out any spray or waves, and the brass strip, if not perfectly water-tight, may be made so by a strip of rubber cloth 3in. wide tacked to both hatches, covering the strip and its joints. Of course neither of these hatches will keep out all water when capsized, but they will be much dryer than the ordinary deck hatches, they cannot be lost or left behind, the cuddy hatch is quickly turned over out of the way, they are easily opened and shut and cannot drop off and allow the contents to fall out if capsized, while being flat, they can be made very strong and will not warp as all curved hatches do.

Now to make the after one water-tight. The coaming inside will be probably 2¼in. deep or a little more, and around its lower edge, as well as across the bulkhead, a beading made of four strips *g, g*, each ¼in. square, is screwed strongly, and on this beading is laid a small tube or band of soft rubber. The inner hatch *c* is a board ⅜in. thick, with two battens on the under side to prevent warping, and is large enough to fit neatly inside the coaming, resting on the rubber tube or washer. To hold this hatch down, a cross beam *e* is used, of oak 1in. square at the middle, where a brass thumbscrew *f* passes through, and 1in. wide by ⅜in. thick at the ends. It is 1in. longer than the distance between the coaming to the bulkhead. This beam slips into two

* In fitting this arrangement the hinges and rubber have been discarded as unnecessary.

notches, one in bulkhead, and one in the coaming at after end of well, at such a height that it can be slipped in freely, when the hatch *c* is in place, when a couple of turns of the thumbscrew *f* brings the hatch down firmly on the rubber. As this inner hatch is a flat board, and is completely protected from sun and water, it cannot warp as exposed hatches do, and it is so covered by the outer hatch, that no water can reach it unless the boat has her masts level with the water. As for simplicity, in spite of the long explanation it is quickly worked, the outer hatch is unhooked and turned over, making a flat table on which to lay articles in packing, the thumbscrew is turned twice, the crossbeam and hatch lifted out, and all is open. The thumbscrew may run into a socket in the hatch, thus attaching the latter and the beam, and a lanyard made fast inside, but long enough to allow the hatch to be lifted off, will prevent either being lost, so that there will be no detached parts.

The objection may be made that the flat hatch is less graceful than the curved one, but on the other hand it can be much stronger, it will not warp, and will certainly be dryer, while folding flat on top, it takes little room when opened. If a tiller is used, it will fit in a socket like the whiffletree fastening and not over a pin. The fore end of the cuddy hatch should have a small beading to prevent any water running into the well. Plate XVIII. shows a view from above, with the outer after hatch opened, side views with the same opened and closed, and a vertical section through the center, with details of beam and thumbscrew and brass-covered joint.

With this division of the boat, the after end will be devoted to bedding, extra clothing and articles which must be kept dry, usually the lighter portion of the load, while forward will be stowed the mess chest, cooking traps, and heavier articles until a proper trim is obtained.

The masts are stepped according to the latest practice of canoeists, and if it were not for the necessity of sometimes unstepping the mainmast while afloat, it would be better to place it 9in. further forward, or 15in. from the bow, and for racing it should be so placed. Both tubes are shown of the same size, 2in. at deck and 1⅛ at bottom, so that the mizzen may be used forward in high winds. The rudder may be of the new drop form, or of ⅜in. mahogany, and will curve quickly aft from the waterline, so that it will not retain weeds or lines which may drift under it.

The forward bulkhead is not shown, as canoeists now differ so much in their ideas as to its proper place. It may be so placed as to leave 7ft. between it and the after bulkhead, being made as tight as possible, or it may be omitted entirely, air tanks being used instead.

The rig for cruising will be about 50 and 20 ft., and for racing, 70ft. in the mainsail and about 25 in mizzen.

The following construction is recommended as being the best, and if properly fastened will be strong and light. Stem and stern, hackmatack knees with proper grain 1in. thick (sided); keel of white oak ⅞in. thick (¼in. outside, ⅜in. for rabbet, and ¼in. inside); width at center 2in. outside and 3in. inside. No keel batten will be needed, the entire rabbet being worked in the keel. The planking will be of clear white cedar ¼in. thick, laid with ⅜in. lap, the lands outside being rounded down at the ends. The upper streak, shown in the plans, will be of mahogany, ⅜in. thick, and should be of strong, tough wood. This streak will be rabbeted on its lower edge, lapping ⅜in. over the streak below. The ribs will be of white oak stave timber ¼x⅜in., spaced 5in. apart, each running across from gunwale to gunwale, except at the extreme ends and abreast the centerboard trunk. They are fastened with copper nails cut off and riveted over burrs, not copper tacks, except at the extreme ends. The weakest point of a canoe, especially those with flat keels, is the middle of the bottom, which in this boat is stiffened by the floor ledges *z z*, Plate II. These, which are placed on each alternate rib, are of oak, or better, hackmatack ⅜in. thick, and deep enough to raise the floor 2 or 2¼in. They will of course be straight on top, where the floor lies, and will fit the rib on the lower side. They are fastened with long, slim copper nails, through the laps and ribs, riveted on the upper side of the ledge. This construction is both stronger and lighter than the use of a thicker keel. The decks will be of mahogany ¼in. thick, and will be screwed to the upper edge of the gunwale, which takes the place of the inner wale and beading, making a strong, light top. The general arrangement of deck frame and coaming has been fully described on pages 52-53.

In finishing the canoe the inside below decks is painted inside of well, and entire outside is varnished, and a gold stripe ⅜in. wide is laid along the mahogany upper streak ¼in. from the lower edge. This gold stripe should be slightly below the surface of the streak, to protect it, a "cove" or groove being ploughed to receive it (Plate XVI.).

PLATE XII.—TANDEM CANOE.

Perhaps no boat taxes more severely the skill of the designer than a modern canoe, as there are so many conflicting qualities to combine in one harmonious whole, within very narrow limits of size, weight and draft, but difficult as the task is with a single canoe, it is still harder with a double boat, and the best that can be expected is a compromise, sacrificing many desirable points to others still more important. Such a canoe should have, first, sufficient displacement to float easily two men of average weight with their stores; secondly, room for both men to sit in comfort, allowing room to move around and stretch the legs; third, room for their stores and clothing; fourth, a foot steering gear by which either can steer. Such a boat is usually intended also to be paddled by one man, if required, to accomplish which no greater length is admissible than 16ft. both on account of handiness and increased weight.

If the boat is intended for a long cruise, where much luggage must be carried, a length of 17ft. with a beam of 32in. would be better, but the same plans may be used, laying down the sections 25in. apart in the working drawing, and placing the moulds at the same distance. To increase the beam the boat may be made 1in. deeper amidships, the heights at stem and stern being the same; then when planked and timbered, but before putting in bulkheads or deck beams, the sides may be sprung apart two inches without affecting the fairness of the lines.

a.	Mainmast tube.	*h.*	Floorboards.
b.	Fore bulkhead.	*i.*	Middle hatch.
c.	Slides for steering gear.	*k.*	After hatch.
d.	Fore hatch.	*l.*	Door in bulkhead.
e e.	Backboards.	*m.*	After bulkhead.
f.	Sliding hatch.	*n.*	Mizzen mast tubes.
g.	Footgear for after man.	*o.*	Rudder lines.

TABLE OF OFFSETS.

	Heights inches.	Half Breadths.					
		Deck.	No. 1. W. L.	L.W.L.	No. 3. W. L.	Diag. 1.	Diag. 2.
I..........	14¼	9⅜	7¾	6⅛	4	7⅞	6⅛
II	12	14	13½	12	9¾	13⅛	11½
III..	11¼	15	14⅝	14⅛	12¼	15	13⅛
IV..........	11	15	14⅜	14⅜	13	15¼	13⅞
V..........	11¾	14	15½	12⅝	10	13⅝	12¼
VI..........	13⅜	9⅝	7¾	7¾	5½	9¼	8¼

Distance from fore side of stem:

To mainmast, 2ft. 9in.
Fore bulkhead, 3ft.
Fore end of well, 5ft.
Sliding bulkhead, 8ft.

To Mizzen bulkhead, 13ft. 6in.
After bulkhead, 12ft.
After bulkhead, 12ft.

Sections 27½in. centers, waterlines 3in. apart, heights measured from rabbet line at midships, planking ¼in. lap of planks ⅝in., timbers $\frac{5}{16}$x⅝in., spaced 6in., keel, stem and stern sided 1in., keelson or keel batten $\frac{5}{16}$x2in., deck ¼in.

A yoke is provided on the afterside of the sliding bulkhead, so that the after man may steer, while another style of foot gear, shown in the body plan, is fitted to the slides forward, which can be used either by the forward man or by a man who is sailing alone. In the latter case the hatch *i* is removed and stowed below, the bulkhead shifted aft to the fore edge of hatch *k*, and the opening at after end of the well closed with a canvas cover. This cover is made of duck, painted, and fits down over the coaming and the edge of the hatch. Around its lower edge are hooks, such as are used on shoes for lacing, and a cord is run through them and over screwheads on the coaming, holding the cover tightly down. A door in the after bulkhead gives room there for storage, the forward compartment being entirely closed.

PLATE XIII.—SAIL PLAN OF TANDEM CANOE.

This sail plan of the double canoe is designed for cruising rather than racing. The area of the mainsail is 63ft., reefing down to 47 and 34ft., and the mizzen is 23ft., reefing to 14ft.

SPARS.

Mainmast—Deck to truck, 10ft.
    Diam. at deck. 2¼in.
    Diam. at truck, 1⅛in.
    Rake ½in. to 1ft.
Main boom 9ft.; diam. 1¼in.
Main yard 7ft.; diam. 1¼in.
Main battens, oval, ⅝×1¼in.

Mizzenmast—Deck to truck, 6ft.
    Diam. at deck, 1½in.
    Diam. at truck, ⅞in.
    Rake ¾in. to 1ft.
Mizzen boom 5ft. 4in.; diam. 1¼in.
Mizzen yard 4ft. 9in.; diam. 1⅛in.
Mizzen battens, oval, ⅝×1in.

SAILS.

Mainsail—Head, 7ft.
    Foot, 9ft.
    Luff, 6ft. 6in.
    Leach, 10ft. 2in.
Tack to peak, 12ft.
Clew to throat, 10ft. 6in.
    a a. Main tack.
    b b. Main halliard.
    c c c. Parrels.
    d. Main sheet sling.
    e. Main sheet.

Mizzen—Head, 4ft. 2in.
    Foot, 5ft. 4in.
    Luff, 4ft.
    Leach, 6ft. 1in.
Tack to peak, 7ft. 2in.
Clew to throat, 6ft. 3in.
    f. Mizzen halliard.
    o. Mizzen tack.
    g. Mizzen sheet.
    n. n. Mizzen toppinglift.
    m. Main toppinglift.
    l l. Main jackstay.

The main tack is led down through a block on the after side of a brass spider band that encircles the mast, and is belayed on the port side nearly amidships. The main halliard leads down through a cheek block on the starboard side of the spider band, and belays on a cleat on starboard side of well, while the downhaul leads through a similar block on the port side of mast to port side of well. The toppinglift is in two parts, fast to the masthead, and leads down on both sides of the sail, and through a bullseye lashed on the underside of the boom. The jackstay is also made fast at the masthead, leads down the port side outside of the sail, and is lashed to the mast just above the boom. In lowering or setting the sail, it lies in the toppinglift and jackstay, which prevent its falling overboard.

The mizzen tack leads direct to a cleat on the deck near the mast, and the halliard leads through a single block lashed to the mast, and is belayed to a cleat near the after end of the well on the starboard side. The mizzen sheet leads to a cleat on the coaming on the port side of the well. The mizzen toppinglift is doubled (on both sides of the sail), and also terminates in crowfeet on the lower ends. The mizzen may be lowered and allowed to hang in it.

Plate XIV.—Canvas Canoe.

Details of canvas canoe building are given on pages 111-114.

Plate XV.—Rowboat.

This boat is of the ordinary type of pleasure boat for rowing and fishing on lakes and rivers. Full details are given on pages 115-122.

Plate XVI.—Rowing and Sailing Boat.

This boat was designed for sailing in a small bay, where it frequently happens that after sailing some distance the wind falls and it is necessary to row home, and it was desired to keep her in a boathouse in order that she might be always dry and ready for use when required.

Her length over all is 13ft., beam 4ft., draft aft when loaded 10in., freeboard 1ft., at bow 1ft. 8in., at stern 1ft. 4in. Owing to her depth, the centerboard, which is rather long, is entirely under the two thwarts, and as much out of the way as it can well be. It is of oak bolted through with $\frac{1}{4}$in. iron, and is fitted with a lifting rod of $\frac{3}{8}$in. brass, with a handle at the top. This rod is so hinged as to turn down on top of the trunk when the board is up, being held by a button. The mast is stepped in a tabernacle so as to be easily removed for rowing. This tabernacle is made of two pieces of oak 3x2$\frac{1}{2}$in. at deck, above which they project 1$\frac{1}{4}$in. At the bottom they are secured to an oak mast step, in which is a mortise for the heel of the mast, and at deck they are let into a piece of board 5in. wide, running athwartship, and screwed firmly to each gunwale. From the mast to the bow a deck of ¼in. mahogany is laid which, with its framing, holds the tabernacle firmly, and prevents any straining of the boat. The forward side of the tabernacle is closed from the step up to within 8in. of the deck, so that the mast will not slip forward when being stepped. The heel is slipped into the tabernacle, the mast raised up, falling into the step, and a brass catch, pivoted at one end, is thrown across the after side at deck and fastened with a turn of the thumb nut shown. The sail is a balance lug, fitted with one batten: Foot, 13ft.; head, 9ft. 6in.; luff, 6ft.; leach, 14ft. 6in.; tack to peak, 15ft.; clew to throat, 13ft. 3in.; batten above boom —2ft. 9in. on luff, 3ft. on leach; mast at deck, 3in.; at head, 1¼in.; mast, heel to truck, 13ft. 8in.

The mast is square in the tabernacle, above which it is round. The head of the sail is cut with a round of 9in., the yard being bent to fit it. The sail is hoisted by a halliard running through a strap on the yard just aft the mast, and hooking into a similar strap forward of the mast. Below it is led through a brass snatch block on the heel of the mast, and aft to a cleat on the trunk, within reach of the helmsman. The tack is spliced to the boom just forward of mast, leads through a bullseye lashed to boom abaft the mast, and down to a cleat on the after side of the mast. The sail may be easily taken from the mast and stowed, for rowing, which cannot be done with a boom and gaff sail. The stem, stern and keel are of white oak, the former two sided 1¼in., the latter sided 4in. outside and moulded 1in. The planking is of white cedar, lapstreak, $\frac{4}{16}$in. thick, the upper streak being of ⅜in. mahogany. The ribs are ⅝x½in., spaced 9in., being jogged down to the plank and copper riveted, the thwarts are of ⅜in. mahogany; rudder 15in. wide, of 1in. mahogany, fitted with tiller and yoke. The gunwales, of oak, are 1x1¼in. at midships and 1x⅜in. at ends. The sides of the trunk, which is covered on top, are of dry white pine, 1¼in.

at bottom and ¼in. at top. They are set flat on the keel, a strip of canton flannel well painted being laid between, and fastened with ¼in. brass screws from outside of keel. The ballast is of gravel, in 30-pound canvas bags.

TABLE OF BREADTHS AND DEPTHS.

		0	1	2	3	4	5	6	7	8	9	10	11	12	13
Depths.	Gunwale to Load Waterline	20	18¼	16⅜	14⅞	13¾	12¾	12¼	12	12	12⅜	12⅞	13¾	14¾	16
	Load Waterline to Rabbet	....	3⅛	7¾	Straight from 3 to 8						9	8½	7½	4¾	...
	Load Waterline to Bottom of Keel	...	7⅛	8⅜	Straight from 3 to 12										10
Half Breadths.	At Deck	.	6¾	12½	17⅛	20⅜	22¾	23¾	24	23⅛	22¾	21½	19⅝	17⅜	15
	Load Waterline	..	3	7⅛	12½	16⅞	20	21⅝	22¼	21⅞	20	17¼	16⅛	9¾	2⅛
	No. 2 W.L.	....	1¾	5⅛	9⅜	13½	16¾	17⅞	19½	19	16⅞	13⅛	7⅜	2⅛	...
	No. 1 W.L.	...	....	2¼	4½	7	9¼	11½	12	11⅝	9⅜	5½	2½	....	...

Plate XVII.—Mohican Sail—Steering Gear.

The members of the Mohican C. C., of Albany, have found the balance lug sail unsuited to their work, river sailing and cruising, and have labored for some time to find something better, the result being the sail now described, devised by Com. Oliver. This sail resembles somewhat the sail of the Alatantis, as made and used by Mr. S. R. Stoddard, but it was fitted by Com. Oliver without any knowledge of the Stoddard sail, from which, however, the idea of the reefing gear was afterwards taken.

In shape the sail is an ordinary balance lug, cut off at the first reef, thus leaving a short luff, and one batten above the boom. The sail is hoisted by a halliard d, which is practically continuous with the downhaul e. The halliard is made fast to a brass ring a on the mast, thence it leads through a snatch block c on the yard, through a block b on masthead, thence through a block m at deck, and returns through a block j, ending in a brass hook. The downhaul e is fast to the batten i, runs down through rings on the sail to brass ring n, lashed to the mast. The two reef lines ff are double, one on each side of the sail, running through block on the boom, and uniting in a single line, which is also part of e, so that the three lines from batten to boom at middle, fore and after ends really run through n, as a single line, the small ring in the bight, into which the halliard hooks, only serving to equalize the pull.

The boom is held to the mast by a brass jaw g, above and below which are leather collars C C, which prevent the boom rising or falling, and render a tack line unnecessary. A parrel may be used on the batten, or a jaw h. The tension on the halliard and reef lines is obtained by the line on block j, by which all is hauled taut.

To set the sail the jaws are placed around the mast (g being between the collars C C), the bight of the halliard, next the ring is slipped into the snatch block c, the downhaul and reef lines e f are passed through ring n, and the end of the halliard hooked into the ring. Now the block

j is drawn aft and its line belayed, putting a tension on the halliard and downhaul. The sail is now ready to hoist. It will be seen that the halliard, with block m, always remains on the mast; in stowing the latter the block j is cast off, leaving the halliard free. To take in a reef, that part of the halliard to which e and f are attached is hauled aft, thus slacking away the other part, and at the same time taking in the reef neatly, with no ends to coil away or belay. It is found in practice that the halliard will slip a little, letting the sail down. To prevent this a little brass cam clutch, k, is screwed to the deck, the halliard d being slipped into it. The roller will jam the cord as it pulls forward, but a pull aft will instantly release it.

Foot	9ft. 6in.
Head	10ft.
Leach	12ft. 6in.
Luff	3ft.
Tack to peak	12ft. 6in.
Clew to throat	9ft. 10in.
Total area	65ft.
Reefed	38½ft.

For description of steering gears, see pages 128-132.

PLATE XVIII.—CANOE HATCHES AND FITTINGS. See page 149.

THE PROGRESS OF CANOEING.

IN the three years that have passed since "Canoe and Boat Building" was first published, the growth of canoeing, as well as other forms of boat sailing, has been very rapid, and the changes in the craft have been many, some marked improvements being made. The principles of building, treated in the first part of this book, are unaltered, and in preparing the present edition the improvements in model, rig and fittings have been described in detail in connection with the best examples of the new canoes, such as Lassie, Pecowsic, Notus and the two new designs.

The year 1886 was a most important one in canoe racing, being marked by the first meeting between the "heavy ballast" English canoes, sailed with crew below, and the various American models with crews seated on deck. Canoe racing was reduced to a science in England some years before it became at all popular in America, the result being that the British canoes were far superior in fitting up and mechanical details to the American craft. As in the case of yachting, the various details of the English canoes have been used and thoroughly tested in this country, with the result of the improvement of some features and the total rejection of others, the leading canoes of the A. C. A. now constituting a distinctly marked national type. The result of the races sailed in New York and on the St. Lawrence in 1886 has been to show that both the "no ballast" canoe and the craft with a moderate amount of ballast, say under 100lbs., is much faster than the Royal C. C. type with upward of 300lbs. One reason for the poor performance of the latter in

American waters appears to be that the type was developed on a narrow river and a small pond, where the wind is very puffy and unsteady; and further, the courses are very short. To meet these requirements, the canoes have relied on a large amount of ballast to carry a lofty sail, necessary to utilize the wind between the banks and to make the canoe safe in the flawy breezes, while owing to the many turns required when a number of rounds of a short course have to be made, the maneuvering powers of the boats were developed to the fullest extent, in fact, so far as to seriously impair the running and reaching. No better example of this can be found than the wonderful working of the Nautilus in Mr. Baden-Powell's hands; with her weight and rockered bottom she turned within her own length, and was as completely under the control of her owner as a bicycle would be. In marked contrast to this were some of the American canoes, which, though far faster off the wind, or even when on a long leg to windward, were slow and uncertain in tacking and maneuvering generally.

Until they were defeated at the A. C. A. meet of 1886, the English canoeists held tenaciously to the inside position, lying down in the boat, and thus were compelled to rely on lead for the necessary stability, but while in America both Nautilus and Pearl were sailed from the deck with the reduction of about 150lbs. of ballast, and in both cases the improvement in speed was most noticeable: in fact if the two had been well sailed from the first in the deck position, with little ballast besides the board, they would have made a very much better showing in the International races. So far as canoes are concerned, it is certain that the day of heavy ballast and displacement is past, and it is equally certain that if the value of the deck position had been understood a few years since in England and the races been sailed over more open courses than the Thames and Hendon Lake, the "heavy ballast" canoe of the Royal C. C. would never have come into existence. Since their return both Mr. Baden-Powell, of the Nautilus, and Mr. Stewart, of the Pearl, have designed and sailed with success canoes of the American type, and with the growth of canoe racing throughout Great Britain that will follow the success of the newly organized British Canoe Association, the canoes are likely to approach that type and the early English canoes of a dozen years ago.

The question of no ballast vs. moderate ballast is by no means so conclusively settled, and though the "no ballast" canoes have won the majority of races in the past two seasons, there is still reason to believe that a moderate amount of ballast is desirable, perhaps part of it being in the form of a centerboard of 40lbs. or under. The ordinary canoe is designed to displace besides her hull, spars, sails and crew, the stores and outfit for a cruise, a weight of about 100lbs. This weight is in a clumsy and bulky form, much of it stowed comparatively high, and no man would carry it simply as ballast for racing, but his canoe is supposed to sail at her best when trimmed with this load for cruising; now when racing, with larger sails, it would seem but proper that the displacement, freeboard and load waterline should be kept as before, loose and clumsy ballast such as blankets and provisions being replaced by shot bags beneath the floor. No other class of vessel is expected to sail exactly as well under two very different conditions, and it is not clear why a canoe should do so. The addition of 100lbs. of ballast does not necessitate appreciably fuller lines, the model may be as sharp and clean cut as in any of the "no ballast" canoes, and though in the past the idea of lead ballast has to a great extent been associated with bulk and full lines, there is still a wide field for the Class B canoeist, especially on open waters, to study and experiment in before giving his order for a Pecowsic.

Looking at the question of model, the examples given cover a wide range, from the light Pecowsic to the new design on Plates XXVII., XXVIII. and XXIX., a canoe of over 500lbs. displacement, and yet of easy form, totally different from the Pearls and the old Nautili. The newest design, a 16×29 racing canoe, Plate XXX., is intended to float a total displacement of less than 275lbs., but the same lines may be utilized in building a 16×30 canoe with 1in. more depth, an addition to the present sheer, thus making a very fast and able canoe for both racing and cruising.

In sails, the fashion has changed to the extent of discarding old rigs without supplying anything specially good to take their place, the endeavor being to get a sail entirely abaft the mast, but at the same time easily reefed or lowered. The most successful effort in this direction is the rig shown in Plate XXIX., devised by Mr. C. J. Stevens, New York C. C. The sails used cn Pecowsic, the invention of Mr. Barney, have been used by him with great success in racing; but the general demand is for a rig that will reef and lower In the West, the sprit sail has been tried on canoes with some success, the sprit being carried down and stepping cn the boom about 4in. from the mast, thus holding the boom down. The lateen, Mohican and balance lug are just now in disfavor, canoeists being engaged in various experiments, and it is impossible to say what the outcome will be. A very good sail is shown on Plate XXVI., that of the Notus, a lowering leg of mutton. These rigs, with many details of fitting, are fully described in connection with the plates.

AMATEUR CANOE BUILDING.

PLATES XIX. and XX—CLASS A CANOE "LASSIE."

The Lassie was designed as an attempt to get good speed close windward work, a fairly light and small canoe to carry a moderate amount of ballast—always a heavy load to han-

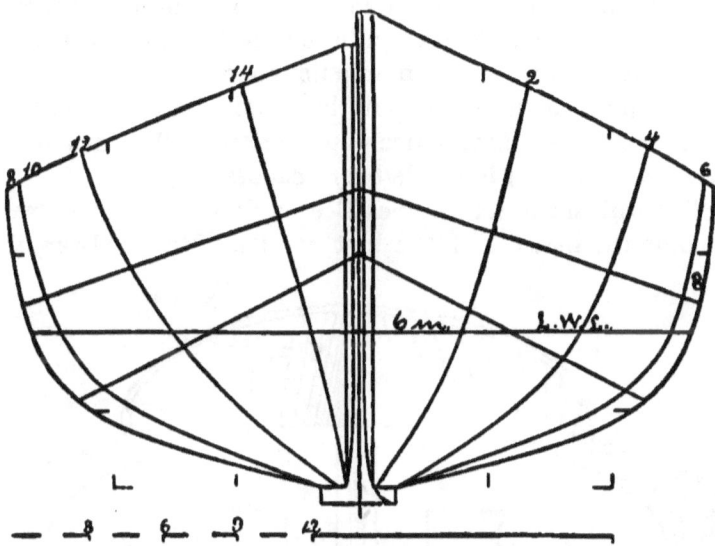

dle—and to be a good cruising canoe for all but very narrow and rapid waters. She has proved herself fast under sail and paddle, easy to handle on the water and ashore, amply large enough for a man of medium height and weight, and needs but 75lbs. of ballast at the most; with all this she is very steady before the wind. She is a Class A canoe, but allowed in Class B races, and just comes within the limits of Class III paddling.

The dimensions—15ft. x 28in.—and the points arrived at were given to Mr. Everson, who worked out the problem in his own way, and to whom all credit for the result is due.

Two flat brass plates were used for the boards, placed as shown in the drawing, as being the best for windward results it was thought, not overlooking convenience at the same time. For cruising the after board can be dispensed with and the slot in the keel plugged. The forward plate can be removed and a wooden board substituted, thus saving about 20lbs. in dead weight. The ballast all goes below the floor, and is held in by the floor boards buttoned down. It is then in the very best place. The canoe is steady before the wind since she draws more water than the Sunbeam—unless heavily ballasted—being narrower. The motion from side to side is a very easy one, quite unlike the quick roll of a flat-floored canoe. The manipulation of the two boards takes time to acquire, so that the maximum result can be obtained. They largely decrease the work that has to be done by the rudder in single board canoes.

The sails made for the Sea Bee—a Sunbeam canoe—were used on the Lassie with the best results. Mr. Tredwen, the

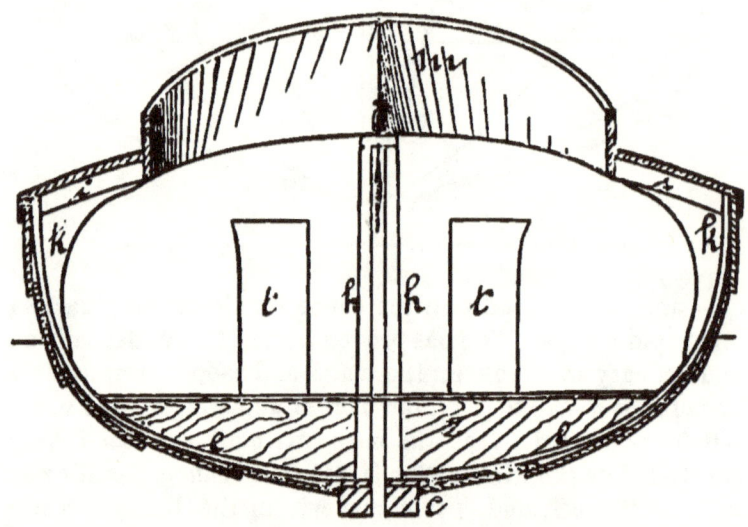

master mind in England on canoe sailing, has warmly commended the short boom and double head gear of this mainsail. The drawing illustrates it clearly. Mast, boom, yard and the two battens are all of exactly the same length, thus making it an easy rig to stow for the spread obtained, 75 sq ft. Very little of the sail is in front of the mast, and the yard peaks up well, both good points for windward sailing. The double purchase at the throat, single at peak, with one halliard, used as a downhaul as well, allows great strain to be put on the yard in hoisting and brings everything as taut as fiddle strings, a flat sail resulting. The topping up of the boom shown in the drawing is not quite true in fact, except when the sail is at rest, or passing over the head of the crew as in tacking or jibing. At other times the pull of the sheets brings the boom end much lower, by the give of the sails, halliards and mast; so much so in close-hauled sailing, when the canoe heels somewhat, that it is about parallel with the plane of the water—the very best position for it to take. The high-pointed coaming, 3in. camber to deck, narrow cockpit (18in.) and flush deck forward make the Lassie a very dry boat at all times. The dimensions are:

Length		15ft.
Beam, extreme		28in.
Depth at gunwale		11½in.
Sheer { Bow		6½in.
Stern		4½in.
Dead rise in 6in		1in.
Crown of deck		3in.
Fore side of stem to—		
Mast tubes, 1ft. 8in., 3ft. ½in.,	11ft.	½in.
Fore trunk, fore end	3ft.	2in.
Fore trunk, after end	5ft.	10in.
Coaming, fore end	4ft.	5in.
Well at deck, fore end	4ft.	10in.
Backrests, r r	7.5in and 9.3in.	
After end of well	10ft.	8½in.
Bulkhead	10ft.	9½in.
Deck tiller	10ft.	10in.
After trunk, fore end	11ft.	2½in.
After trunk, after end	12ft.	6in.
Deck hatch, fore side	11ft.	11in.
Deck hatch, after side	12ft.	9in.
Width of cockpit	1ft.	6in.
Coaming, height at fore end		3in.
Coaming, height at middle		2in.

Waterlines 3in. apart; buttock and bowlines, 5in. apart; sections, 2ft. apart, from fore side of stem; floor above garboards, 8½in.; stem and stern sided 1in., keel sided (width) 3in; moulded (thickness) 1in.; keel batten, ¼in.×4in. at amidships; siding, ¼in.; ribs, ¼×5-16, spaced 5in., 9in. at ends; deck, ¼in.; diameter of mast tubes, 2in; floors, 5-16in.

REFERENCES.

- *a*, stem, hackmatack.
- *b*, stern, hackmatack.
- *c*, keel, oak.
- *d*, keel ba'ten, oak.
- *e*, ribs, oak.
- *f*, bulkhead, pine, ½in.
- *g*, headledges, oak.
- *h*, sides of trunk, pine ½in.
- *i*, deck beams, pine.
- *k*, knees, oak or hackmat'k.
- *l*, maststeps, oak.
- *m*, coaming, oak.
- *n*, hatch, mahogany.
- *o*, after hatch.
- *p*, deck hatch.
- *q*, center strip of deck, mahog.
- *r*, back rests, oak.
- *s*, heel brace, oak.
- *t*, steering pedals, oak.
- *u*, deck tiller.
- *v*, centerboard hinges, brass.
- *z*, floor ledges, cedar.

The keel batten, ¼in. thick, is a separate piece; but it would be better if worked in one with the keel. The centerboard trunks are both below deck, closed on top and opening only on the bottom. The boards, of sheet brass, are hinged by means of two L-shaped pieces, shown at *v*, one on each side of the board. These pieces are each fastened to the keel by a screw from the outside. To remove the board the canoe is turned over, the two screws taken out, and the boat turned back, when the board will drop out. The fore board is of $\frac{5}{16}$in. brass, weighing 15lbs. The after board is of ¼in. brass, weighing 5lbs., and is quadrant-shaped. Each is filed to a sharp edge. They are hoisted by cords, the forward pendant belaying on a cleat on after end of the trunk; the after pendant coming through the deck to a cleat on starboard side deck, abreast the canoeist. The three mast tubes are of uniform size, 2in. inside, so the masts may be interchanged. The rudder is of ⅜in. oak, thinned down at the edges, the yoke being a semi-circular piece with a score in it for the rudder lines, of brass chain. The foot gear consists of two oak pedals, *t t*, fitted to the floor boards with brass spring hinges. When two are paddling, the after man uses the braces, *s s*, in the floor, the back-board for the forward man being at *r*. The hatch, *o*, is made with an outside rim, fitting over the coaming and close to the deck. There is no fore bulkhead, as usually fitted, and the sliding bulkhead is also omitted, a piece, *r*, taking its place in supporting the hatch and carrying the back-board. The broken lines in the sectional view show the inner edges of the planking, the widths of the same at midships being given in the cross section, page 165.

PLATE XXI.—CLASS B CANOE "SUNBEAM."

This canoe was built early in 1885 by James Everson for Mr. J. F. Newman for a cruising canoe. The model has since become very popular and a number of these boats were present at the meet of 1885. The boat is intended for general use in wide waters where ballast is desirable, and upward of 100lbs. of shot in bags is carried. For use in narrower waters a flatter floor would be needed, no ballast being carried. Though intended for a cruiser, the canoe has proved very fast and several of the same model have taken a place among the racers of their class. The interior arrangements are of the usual form. At d, e and f are bulkheads, that at f being movable. The well is covered with hatches, in the usual style, $a\,a$ and $c\,c$ are airtanks of phosphor bronze. The dimensions are:

Length	15ft.	
Beam	2ft.	6in.
Depth amidships		11in.
Sheer at bow 7in., at stern		5½in.
Crown of deck		2in.
Distance from foreside of stem to—		
Mainmast	1ft.	8in.
Bulkhead	4ft.	3in.
Fore end of well	4ft.	10in.
After end of well	10ft.	0in.
Sliding bulkhead	9ft.	6in.
Mizzenmast	11ft.	8in.
After bulkhead	11ft.	6in.

The dimensions of frame, planking, etc., are the same as in the Lassie. The rig of the Sunbeam is two balance lugs of 70 and 35ft. for racing, or 50 and 15ft. for cruising.

In the fall of 1886 the afterboard and trunk were removed, a brass drop rudder was added, and the deadwood at stem and stern cut away as far as possible, the canoe being too slow in stays.

PLATES XXII. AND XXIII.—PECOWSIC.

Perhaps the greatest pleasure that comes to most owners of boats is not so much in actually possessing the fastest craft of all, as in the continual effort to gain that desired end by surpassing the similar efforts of others: a competition that is often more exciting and engrossing than the final test by which in a few hours the results of this labor are proved to be satisfactory or the reverse. It often happens that the development of the highest capabilities of a yacht is the work of several seasons of careful and painstaking effort, and of many changes and experiments; work that calls into play all the inventive faculties and reasoning powers, but that ultimately brings a far higher reward when success is attained than do the briefer and less intellectual struggles of the regatta course. It is from this point of view that the canoe, looked on contemptuously by many as a mere toy, and unworthy of serious notice, commends itself to a large number of intelligent men as a fitting subject for their study. Where the first cost of a yacht may range from ten to thirty thousand dollars, and the cost of any changes are in proportion, there are comparatively few who are able to follow the sport to its fullest extent; but in the canoe, while the cost of boat, outfit, and a season's racing will not exceed two or three hundred dollars, the interest is no less intense, the competition is as keen, and the rewards are great in proportion. In no other sailboat, perhaps, is there so much room for ingenuity and invention. The small size of the boat and the amount of work her crew of one must do make it necessary that everything should be arranged to the best advantage, while the strong competition between the various craft, both in home and distant races, is a constant stimulus toward improvement in model, rig and fittings. Every one familiar with the leading canoes will recognize the fact that each testifies not only to her owner's skill as a sailor but also as a designer, rigger and inventor, and that each boat, while all are alike to the casual observer, possesses a marked individuality of its own.

From this point of view no less than from the prominence which he has lately attained, the canoeing experience of Mr. E. H. Barney of the Springfield C. C., is a most interesting and instructive one. Taking up canoeing as a novice, at an age when most men have given up such sports, he has in less than three years won a most enviable place among the leaders of the sport. Mr. Barney began his canoeing early in 1884 with a lateen rigged canoe of good model and fitted in the best style of the leading builders; but a short trial served to show many points that were capable of improvement. The rudder, fitted in the usual manner, was not perfect by any means and soon gave place to an original method of his own that is no less admirable for its effectiveness than for its simplicity and mechanical perfection. With this came a new deck tiller, a rudder yoke that could not foul the mizzen sheet, the "fishtail" rudder and many smaller details. A little experience brought changes in sails, rigging and centerboard, until this novice was soon looked upon as one of the leaders in the field of canoe inventions. His first boat was soon discarded for a better, and this in turn made way for a third, until the fifth, the well-known Pecowsic, was purchased last year.

Three of these canoes have borne the name Pecowsic, the one here illustrated being the third; and this, like its predecessor, was built for Mr. Barney by F. Joyner, of Glens Falls, N. Y. The model was made by the builder to Mr. Barney's order, and the method of construction, the smooth-skin lap, is the same as Mr. Joyner has employed so successfully for some time. The general arrangements, the positions of masts, boards, etc., as well as the entire rig, were planned by Mr. Barney. The accompanying lines were very carefully taken from a small drawing, and the full-sized outlines of the moulds, furnished by Mr. Joyner; but some fairing was necessary to put the drawing in its present shape. The midship section shows far less deadrise than an inspection of the boat itself would indicate, the cutting away of the ends giving an idea of a sharp V section, quite different from what the drawing shows. The bulk of the hull is small, and its internal capacity limited, though it is claimed that there is ample room for cruising outfit, and that the boat is well adapted for general work. There is but one bulkhead just abaft the well, closed with one of Joyner's circular hatches. The fore end of the boat is entirely open, to permit of the stowage of spars and sails. The board is a sheet of thin brass only 30in. long but rising high above the top of the low trunk, shown by the dotted lines in the sheer plan; before the Meet of 1886 the board was shifted 10in. forward of the position shown. The well is short and far aft, while the trunk interferes with the room, and sleeping is difficult if not absolutely impossible. As no ballast is carried and there is little weight of metal, the danger of sinking if filled is removed, especially as one air tank is carried in the after end. The weight of the hull, about 100lbs., is nearly all made up of wood. The dimensions are as follows:

Length, extreme	15ft. 10¼in.
l.w.l	15ft. 6 in.
Beam, extreme	28⁶in.
l.w.l	27¼in.
Depth, amidship	9¼in.
Sheer, bow	8 in.
stern	5¼in.
Draft	6²in.
Crown of deck	2¼in.
Well, width	17 in.
length	5ft. 00¼in.
Foreside of stem to foremast	7 in.
mainmast	6ft. 6¼in.
mizzenmast	13ft. 4 in.
well, fore end	6ft. 10¼in.
well, after end	12ft.
bulkhead	12ft.
centerb'd trunk, fore end	6ft. 7 in.
centerb'd trunk, after end	9ft. 1 in.

The first station is 2ft. from stem, the others are each 18in. apart. The waterlines are 2¼in. apart.

The most peculiar feature of the boat is her rig, differing as it does from anything else in canoeing. The advantages of the simple leg o' mutton sail were too apparent not to be quickly seized upon by canoeists, but a difficulty was experienced in obtaining sufficient area; besides which the sail is hard to hoist in such small sizes, the mast rings having no weight and being liable to jam very frequently. After being used for some years the sail was abandoned; but after trying the others in turn, Mr. Barney was attracted by the simplicity and efficiency of the leg o' mutton sail, especially for an unballasted boat of narrow beam, and began to experiment with it, making his own sails. To overcome the first objection, he added a third sail, thus making up the area; while it was well distributed over the length of the boat, and at the same time the center of effort is kept low, an important point in such a craft. The second difficulty, that of handling, was disposed of by lacing each sail to its mast and not attempting to hoist or lower it, the mast and all being removed and a smaller substituted if reefing was required. To do this successfully, five sails are carried, the masts and tubes being all of one size. Three of the sails must be set at once, the other two being stowed below. It would seem that not only is this shifting a matter of difficulty in many cases, but that the sails below would be a serious incumbrance in so small a boat, but those who have used her state to the contrary.

The sails and spars are as follows:

Mast.	Boom.	Battens, No. of.	Area.
No. 1, 8ft.	5ft. 10in.	1	22 sq. ft.
No. 2, 10ft.	5ft. 10in.	2	28 sq. ft.
No. 3, 10ft.	5ft. 10in.	2	33 sq. ft.
No. 4, 10ft.	5ft. 10in.	2	38 sq. ft.
No. 5, 11ft. 8in.	5ft. 10in.	3	42 sq. ft.

The sails shown in Plate XXIII. are Nos. 5, 4 and 2, No. 1 being indicated by the dotted lines, while No. 3 is similar to No. 4, but smaller. The greatest possible area is 113ft., the least 22ft. The booms are limited in length by the distance between main and foremasts, and the first batten in each sail, except No. 3, is to gain more area. The other battens were added to make the sails sit properly, as they bagged in places. Mr. Barney has used very light spars, the masts being slender sticks with a quick taper, and so having little weight aloft. The booms are fitted with brass jaws (Plate XXIX.a), allowing them to top up easily. The sails are fastened to the spars by small wire staples, such as are used for blind slats. No lines of any kind are used except the sheets, and the extreme limit of lightness and simplicity is reached.

Since Pecowsic's success in 1886, a number of similar craft have appeared in the races; some of them much fairer in model than the lines here shown, though all by the same builder. None, however, have equalled the record of Mr. Barney's boat, which is good evidence that the reason for Pecowsic's speed, which has puzzled so many canoeists, is to be found not so much in her model, as in the skill and care with which she is rigged, fitted up and handled. In 1887 Pecowsic was sailed by Mr. Geo. M. Barney, son of her owner, the latter using a new canoe of similar model, but rockered up much more aft, named Lacowsic. She was 15ft. 10in.×27¼in., built at Springfield under Mr. Barney's supervision, with a double skin. The sails were identical with Pecowsic's. Both of these canoes made an excellent showing in the season of 1887.

PLATES XXIII., XXIV., XXXa.—"NO BALLAST" CANOE VESPER.

The utility of some ballast and of boats built to carry it is generally admitted when open-water sailing is in question, but there are some locations where a totally different type of boat has come into use, and has found great favor at the expense of the heavier-ballasted craft. This has been the case particularly at Albany, where canoeing is

AMATEUR CANOE BUILDING.

confined to the Hudson River, with occasional excursions to neighboring streams of a similar character. The boats first used by the Mohican C. C. were of the Shadow and similar models as built a few years since, but for some years the club has displayed great activity in the hunt for improvement, and besides the sail and fittings generally known by their name, they have devoted much attention to the question of model. Vesper was designed by Mr. R. W. Gibson in 1885–6, and built by Mr. J. H. Rushton, the hull being lapstrake and very lightly built.

The table of offsets is as follows:

Stations.	Heights.		Half Breadths.							
	Deck	Rab't	Deck	10in.	6½in.	4½in.	2½in.	1in.	Keel.	Ding. a.b.c.
	Ft.In	Ft.In	Ft.In	Ft.In	Ft.In	Ft.In	Ft.In	Ft.In	Ft.In	Ft.In
0	$1 8^4$	0	0^1	......	......	......	......	......	0^1	0^1
1	15	0^4	6^1	5^5	4^6	4^1	2^7	1^2	0^4	5^5
2	12^4	0^2	10^5	10^2	9	8^2	6^7	4^6	1	9^6
3	11	0	13^4	13^3	12^3	11^6	10^2	8^3	1^2	12^4
4	10^2	......	15	15	14^4	14^1	13	11^3	1^4	14^7
5	10	......	15^2	15^2	15^2	15	14^2	12^6	1^4	15^7
6	10^2	......	14^6	14^6	14^3	13^7	13	10^7	1^3	14^1
7	11^3	0^1	12^3	12^2	11^4	10^4	8^6	5^7	1^1	11^4
8	13^2	0^2	7^5	7^1	6	5	3^3	1^3	0^7	6^0
9	16	0^4	0^2	......	......	......	......	......	0^2	0^2
Foreside stem from station 0			0	0	0^1	3^4	7^1	14	......	......

The dimensions are:

    Length over all..................................15ft. 6in.
    Beam, extreme.................................. 30¼in.
        l.w.l............ .... 30 in.
    Draft, excluding keelson...................... 4¼in.
        including keelson.... .... 5 in.

Freeboard, bow	14 in.
amidships	5¼in.
stern	11¼in.
Sheer, bow	8¼in.
stern	6 in.
Rake, sternpost	2 in.
Crown of deck	2 in.
Diameter of mast tubes	1¾in., 1⅞in., tapered to about 1in.

The rig shown in Plate XXIII., has the ordinary Mohican sails, rigged as shown in detail in Plate XVII.

PLATES XXV. AND XXVI.—NOTUS—RACING AND CRUISING CANOE.

After a season's use of Vesper, Mr. Gibson sold her and made a new design embodying some improvements, and in 1887 Notus was built. She is a 16×30 canoe, much like Vesper, her 6in. waterline being the same; but she is cut away more under water forward, giving a slightly hollow entrance, to improve her performance in rough water. The same long, fine bow and full stern already tested by Mr. Gibson have been retained, but the extremely broad and long floor is modified. Notus having about 10in. flat and an elliptical form of midship section, leading into the turn of the bilge, which gives remarkable strength. The stability is not perceptibly diminished by this slight rounding, and it probably assists turning, which Notus does with the greatest case. The canoe was built by Charles Piepenbrink, of Albany, under Mr. Gibson's personal supervision, and is a remarkably fine piece of work. She is a smoothskin, with only three planks to a side, the ribs spaced 6in. and fastened with brass screws from the inside. A few screws were required from the outside, but they are 12in. apart, leaving the bottom absolutely smooth. The planking and decks are of white pine, and the trimmings of maple and mahogany, two narrow beads along each side. The board is of sheet brass, $30 \times 13 \times \frac{1}{16}$in., dropping through a low trunk. There are four bulkheads, with a low hatch in the fore deck.

The table of offsets is as follows:

Stations.	Heights.		Half-Breadths.					
	Rabbet	Deck.	Deck.	8in.	6in.	4in.	2in.	Keel.
0..	....	13	0^1	0^1	....	....	....	....
1..	1	15^6	3^7	2^5	2	1^3	0^6	0^2
2..	0^8	14^3	7^7	5^3	4^3	3^2	1^7	0^4
3..	0^2	12^7	9^7	8^2	7	5^3	3^2	0^7
4..	0	11^6	11^7	10^5	9^3	7^6	5^1	1^1
5..	0	10^7	13^4	12^5	11^6	9^7	7^2	1^4
6..	0	10^3	14^3	13^7	13^1	11^6	9^2	1^4
7..	0	10	14^6	14^3	14^2	13^1	10^7	1^4
8..	0	10	14^7	14^7	14^4	13^7	12	1^4
9..	0	10	15	15	14^7	14^1	12^3	1^4
10..	0	10^1	14^7	14^7	14^6	13^7	12^1	1^4
11..	0	10^2	14^4	14^3	14^1	13^1	11^2	1^4
12..	0	10^6	13^4	13^2	12^6	11^6	9^4	1^3
13..	0^1	11^4	11^4	11^1	10^4	9^3	7	1^1
14..	0^2	12^5	8^3	7^7	7^2	6^2	4^2	1
15..	0^6	14^2	4^3	4	3^3	2^6	1^3	0^4
16..	....	16	0^1	0^1	0^1	0^1	0^1	0^1

The dimensions of Notus are:

Length over all..............................16ft.
Beam..30in.
Depth...10in.
Sheer at bow...................................8in.
    at stern....................................6in.
Fore side of stem to bulkheads,
        2ft. 6in., 5ft., 10ft. 6in., 13ft. 6in.
        mainmast.................9in.

Fore side of stem to mizenmast..............11ft. 10in.
board, fore end..........5ft. 8in.
after end..........8ft. 2in.
coaming, fore end..........4ft. 2in.
after end..........11ft. 6in.

The sails were devised by Mr. Gibson, and are made of bleached muslin in one width, the edges being bound with wide tape. The battens fit in pockets in the usual manner. The spars are very light, the masts 2in. square at deck, tapering to ⅝in. diameter at head, the main boom 1¼in. diameter, battens ½in. thick. The dimensions of sails and spars are:

	Main.	Mizen.
Mast, deck to head	13ft. 9in.	11ft. 4in.
Sail, on foot	9ft.	6ft. 9in.
along first batten	8ft. 8in.	6ft. 6in.
along second batten	8ft.	
luff	13ft.	10ft. 6in.
leech, total	14ft. 7in.	11ft. 8in.
leech, above batten	10ft. 3in.	9ft. 6in.
spacing of battens, fore end, 1 t.	11in.	1ft. 11in.
after end,	2ft.1½in.	2ft.1½in.
area, square feet	69ft.	39ft.

The first reef leaves 52ft. in main and 26ft. in mizen, the second reef leaves 35ft. in main. The mizen can be stepped forward and a storm mizen added. The luff of the mainsail is roached 4in. in 13ft. and the luff of the mizen in the same proportion. The usual reef gear is added. The sails are hoisted by halliards and lowered with downhauls, the attachment to the mast being by a lacing, as shown. This lacing (Plate XXIX.*a*) is similar to the ordinary hammock or netting stitch, the loop or mesh loosening as soon as the halliard is cast off, but as the latter is hauled taut the meshes lengthen and draw the luff closely to the mast. It has been suggested that a few small beads on the lacing near each knot would make the sail run easier in hoisting and lowering. In the fall of 1887 Notus was sold to Mr. R. W. Bailey, Pittsburgh C. C.

PLATES XXVII., XXVIII. AND XXIX., 15×31, CLASS B, RACING AND CRUISING CANOE.

This design was made in 1883 for Mr. Wm. Whitlock, of New York, by Mr. John Hyslop, and from it the well-known Guenn was built in the winter of '83-4. Unfortunately she was too large for the A. C. A. limits, and in order to race she had to be shortened, drawn in and re-decked after being in use for some time, which altered the lines materially. Further than this, she was heavily built, with a large keel and a fan centerboard, thus handicapping her in racing. In spite of these disadvantages she has shown good speed at times, and there is every reason to believe that the model is a fast one, though not fairly tested in Guenn.

In the present design the outlines of the hull have been preserved intact, but the raking sternpost of Guenn has been replaced by a straight one, in accordance with the latest practice, and the exterior keel and the deep deadwood aft have also been cut away. The scantling is given for a light but strong cruising boat, and with the model and suitable rig she should prove a winner in the races as well. A canoe built closely to these lines would be a very different craft from the old Guenn. The hull is large and powerful and well fitted to carry a large load of stores and gear, or heavy board and some ballast for racing. Her place should be about New York and on broad waters, rather than on the upper Hudson and the Connecticut; and she will undoubtedly make an excellent all-round boat and an able racer as well, even though she should prove unable to master the Pecowsics in all weathers. The design is the first that Mr. Hyslop had ever made for a canoe, a class of boat with which he was not familiar, and the dimensions and the leading particulars were given by Mr. Whitlock, the designer being responsible only for the lines of the model.

It will be noticed that the drawing measures but $30\frac{3}{8}$in.

extreme beam, with planking. If the moulds are carefully made to this size the boat when planked may be allowed to spread a little, bringing her to 31in., leaving ¼in. inside the A. C. A. limit before the deck is put on. This is better than building to the exact width, as the boat will always spread a little. At the same time it would be possible to build a 30in. boat from the same moulds, using a little care in drawing the sides together before timbering, and fastening them well until the deck frame is in, but it is always best to build a light boat narrower rather than wider than she is to be, and then allow her to spread a little. If a smaller boat is desired the design may be cut down in depth, taking 2in. off the sheer all around without impairing its integrity. Such a boat would of course need no ballast, and would be a very fair match for Notus, Vesper and others of that class. The line shown for crown of deck is simply drawn in with a batten to make a fair sweep, with a crown of 2in. at midships. This will not allow one beam mould to be used throughout, as the round of the deck beams varies at each station, but it will make a handsome deck. The dimensions are:

  Length, extreme.................. ....15ft.
  Beam, extreme .... 2ft. 7in.
  Depth, amidships....... .... 1ft. 0¼in.
  Sheer, bow............................. 6⁵in.
     stern............................ 2⁴in.
  Draft, including keel..................... 6²in.
  Displacement, to above draft............ 535lbs.
     Per inch immersion...... 130lbs.
  Area, midship section.................... .965 sq. ft.
    loadline plane..................... 23.88 sq. ft.
    lateral plane..................../..... 7.33 sq. ft.
  C. L. R. from foreside of stem........... 7ft. 0⁴in.
  Waterlines, 2in. apart; stations, 1ft. apart.

The question of construction is still as much in dispute as ever, and with little probability of a final settlement, as each of the leading methods has its strong points, together with some marked disadvantages. The large and increasing demand for canoes of all sizes has stimulated

AMATEUR CANOE BUILDING.

TABLE OF OFFSETS.

Stations	Heights		Half Breadths						
	Deck	Rabbet	Deck	No. 10	No. 8	No. 6	No. 4	No. 2	Rabbet
0..	1 7^1	......	0^1	0^1	0^1	......	......	......	......
1..	1 5^6	1^3	3^7	3^1	2^5	2^1	1^3	0^6	0^4
2..	1 4^3	0^3	7^2	6^2	5^4	4^4	3^3	2	0^5
3..	1 3^2	0^1	9^7	9^1	8^4	7^2	5^6	3^4	0^7
4..	1 2^4	......	11^7	11^4	11	10^1	8^3	5^3	1^1
5..	1 1^6	......	1 1^3	1 1^2	1 0^7	1 0^2	10^4	7^1	1^2
6..	1 1^2	......	1 2^2	1 2^3	1 2^1	1 1^6	1 0^2	8^7	1^3
7..	1 0^6	......	1 2^7	1 3	1 2^7	1 2^4	1 1^4	10^3	1^4
8..	1 0^5	......	1 3^1	1 3^2	1 3^1	1 2^7	1 2^1	11^1	1^4
9..	1 0^4	......	1 3^1	1 3^2	1 3^1	1 2^7	1 2	10^7	1^3
10..	1 0^4	......	1 2^5	1 2^6	1 2^5	1 2^2	1 1	9^3	1^2
11..	1 0^4	......	1 1^6	1 1^6	1 1^4	1 0^7	11^1	7^2	1^1
12..	1 0^7	0^1	1	11^4	11^2	10^6	8^6	5^1	0^7
13..	1 1^3	0^1	9^3	9	8^4	7^4	5^5	3	0^6
14..	1 2^1	1^3	5^5	5	4^4	3^5	2^5	1^2	0^4
15..	1 3	......	0^1	0^1	0^1	0^1	0^1	......	0^1

the inventive powers of builders, both amateur and professional, with the result that many new methods have of late been tried with more or less success. The first American canoes were all lapstreak, and when, in 1881, the author first introduced the ribband-carvel method of building a smooth-skin boat, then used in England, it met with no favor from American canoeists, there being a strong prejudice in favor of the lapstreak. Fashions change in canoes as in dress, and for the past two years smooth-skin boats have been the rage, perhaps for no better reasons than those once urged against them. There are to-day half a dozen excellent methods of construction from which the canoeist can choose with a fair certainty of having a first-class canoe, and it would be a very difficult matter for an unprejudiced judge to say which, if any, is absolutely the best. After some experience in building and using canoes, and some familiarity with the different methods of construction, we feel safe in recommending the lapstreak, if properly built, as the best for cruising. Whether she will prove the fastest alongside of some of the smooth-hulled racers is still an open question, and most canoeists would say "No" to it, but some badly built lapstreaks have done so well in the races at times that there is every reason to think that an absolutely smooth skin counts for little against fine and well laid laps.

Of lapstreak work there are all kinds, from the clumsily-built pram of the Norwegians, with wooden plugs in place of nails, to the carefully planked canoe with a land something like that shown at 3. The edges of both plank must be very accurately beveled, the outer edge to a thickness of $\frac{1}{16}$in. or a little more, being rounded off as shown in sandpapering. The common lap is shown in 2, a strong joint, but giving a poor surface. The strength of the lapstreak has repeatedly been proved beyond question, it will stand both wear and hard knocks, while it is very light. With the requisite care and skill the bottom of the boat may be made to compare very favorably with any of the smooth-skin methods, and after a season of rough cruising the lapstreak canoe will probably be in better condition than the others. One method of securing a smooth skin is shown in 1, the plank being cut with a special plane, taking half out of each. The ribband carvel canoes built by the author in 1881 and 2 had a strip inside the seam, as in 4, both planks being nailed to the strip, the ribs were put in afterward, being jogged over the ribband. The Albany canoes are built now on a similar plan, but by a method hardly suitable for amateur work. A strong mould or last is first constructed of the shape of the inside of the canoe. The deadwood and all the ribbands are fastened to this last, the ribbands are cut away so that the ribs can be let in flush, then the planks are laid and screwed to the ribbands and ribs. The construction of the mould or last over which the canoe is built is almost as troublesome and costly as the canoe itself, so this method is only practicable when a number of canoes are to be built of the same model. In the method shown in 4 the usual plank moulds are used, as in lapstreak work, so the process is well adapted to the needs of the amateur.

The details of board, rudder and steering gear here given were not part of the original Guenn, but are all original with the author. They are unpatented, and at the service of all.

In the construction of a canoe of this size, presumably to carry some ballast, the following scantling will give strength with little weight. The stem and sternpost will be of hackmatack, sided so as to end all lines fairly at the extreme ends. It has been the custom to make these pieces of 1in. stuff, which in most cases makes an angle in the waterlines at the rabbet in stem and stern. To avoid this they must be sided as shown by the full size plan which will be laid down before building. The stem will be ½in. on its fore edge, the sternpost ⅜in. The stem band will be made from ⅜in. half round brass rod, and after it is screwed in place the wood and brass will be filed down together until the lines are carried out straight and fair. The usual method is shown at 5, the proper one at 6, the dotted line in the latter shows where the stem is left a little thick in cutting the rabbet, being filed and planed down after the planking is completed and the stem band in place.

The keel is usually made at least 1in. thick, but this is in no way necessary. It is better to make it as thin as possible and quite wide. In the present case it is shown 3in. wide and ¾in. thick, but it might well be 4 or 5in. wide outside and but ½in. thick. The stiffness of the bottom depends but little on the thickness of the keel, the main point being to brace the whole floor system thoroughly by the ledges which carry the floorboards. The keel may be regarded simply as a wide bottom plank, and so need be but little thicker than the other planks. It is here shown $\frac{3}{4}$in. thick, or $\frac{1}{2}$ inside of boat, forming the rabbet, $\frac{1}{4}$ for thickness of garboard, and $\frac{1}{4}$ projecting outside of garboards. Really all that is needed is $\frac{1}{2}$ for rabbet inside and $\frac{1}{4}$ for garboard, leaving no projection outside. If at the same time the keel be 5in. wide it makes a wide, flat surface on which the canoe will rest when ashore or on which she may be easily dragged over rocks or logs. It often happens that while a canoe can be dragged easily enough, there is great difficulty in holding her upright on her keel and at the same time dragging her, but with a wide keel she will always keep upright with little aid. In a canoe with much deadrise the keel, if very wide, may be slightly rounded, and in any case it should be protected by $\frac{3}{8}$in. strips of flat brass $\frac{1}{16}$in. thick, screwed to the entire length and soldered or brazed to the stem band. With such protection and a similar strip 4ft. long on each bilge, the boat will stand safely a great amount of hard work, and as for racing, the lessened damage to the planking will more than compensate for any slight friction of the brass. The keel should be worked from an inch board, leaving the full thickness at the ends to aid in forming the deadwood, but planing down to $\frac{3}{4}$in. or a little more at the middle half of the boat, say for a length of 8ft. The scarfs of stem and sternpost as well as the fastenings are shown. It is now the fashion to place the mainmast as far forward as possible, to do which the step must come in a part where the lines are very fine, which necessitates a very thick step. The one shown is of clear white pine, to save weight, and is fastened in when the frame is put together. It must be very securely bolted and must be trimmed carefully to shape just as the deadwoods are, so that the planks will fit accurately against its sides. The step for the mizen may be put in afterward, in the usual way.

A simpler method of building the trunk than that on page 91 is to put the sides of the case together with the headledges between, riveting them up, then to get out a piece of pine as wide as the top of the keel and ½in. thick, which piece is screwed fast with long screws to the bottom of the trunk. The lower side of this bedpiece is then accurately fitted to the top of keel and screwed fast by short screws passing through it into the keel. The labor of rabbeting is avoided and the joints, if laid with white lead and Canton flannel, will be perfectly tight. The sides of the case will be of clear white pine, ½in. on lower edge and ⅜in. at top. The headledges will be 1in. wide and thick enough to allow ample room for the board. The planking will be ¼in., of white cedar. The upper strake, of the width shown in the drawing, will be full ⅜in. thick, being rabbeted on the lower edge so as to lap over the strake below, showing ½in. outside. There will be no gunwale, the deck being screwed to this upper strake. The ribs may be a little less than ⅜×¼in., spaced 6in., with two rivets between. The bulkheads will be ⅜in., of clear white pine. The deck will be of ¼in. mahogany. The ledges for the floor will be of spruce or hackmatack, as deep as the distance from floor to garboard, and ⅜in. thick. They should be very securely fastened with long rivets through each lap and the keel, at least two through the latter. If long nails cannot be had, copper wire can be used, with large burrs on each end. These ledges should be placed alongside the trunk, of course being in two pieces each, and a stout one should be placed against the afterside of the trunk and screwed to the headledge. If well fastened they will make the bottom perfectly rigid, no matter how thin the keel may be, while boats are often found with a keel 1¼in. thick that will constantly work under the strain of the board or when ashore. The centerboard trunk itself plays a very important part in strengthening the hull, if properly built into the boat and coming up to the deck.

The shifting bulkhead is placed with a slight slant, to

accommodate the back better than if vertical. In planning the arrangements, every effort has been made to give plenty of room for cruising, not a mere hole where a man may stow himself for half an hour, but room to carry stores and bedding, to cook, to change one's clothes and to sleep in comfort. At the same time the sailing qualities could not be neglected, and the board has been placed as far aft as possible, with a provision for shifting it still further aft in sailing, as will be described later. The distance between bulkheads is 7ft. 5in., and as the boat is both wide and deep, this space should give plenty of room for all stores; but the after compartment might be fitted with a hatch if more room were desired. As now arranged, the mess chest could fit on one side of the trunk, and the clothes bag on the other; while the tent and the bedding, the latter tightly strapped in a waterproof bag, would be in the locker under the movable hatch. In sleeping, there would be a clear length of 5ft. $5\frac{1}{4}$in. from bulkhead to after end of trunk, and the feet could extend for a few inches into the space beside the trunk lately occupied by the clothes bag, now doing duty as a pillow. The tent, of course, would be set, the bedding spread, and the hatch and bulkhead removed for the night. The space under the side decks affords plenty of room for oilers, spare gear, apron, etc. The well is large, as in summer cruising a man requires plenty of room, and if the feet and legs are kept continually below deck they will be very warm. Sleeping, cooking, and the positions one naturally takes to obtain a rest when afloat all day, demand more room than some racing men seem to think necessary. To close the large well in stormy weather, the best plan is an apron of heavy drill, fitting over the pointed coaming and tightly laced along the sides to screw heads outside the coamings. The apron may extend as far aft as the cleats shown. The mast tubes are $2\frac{1}{4}$in. at deck, tapering. Both are of the same depth, so that the masts may be interchangable. It will be noticed that the coaming is cut down very low at the after end. It was formerly the custom to make the coaming as high there as anywhere, but this is not necessary, as little water will come aboard in the center of the deck, and by cutting the coaming down low the need of raising the deck tiller is avoided.

The arrangement of the centerboard is peculiar and entirely novel. The usual arrangement has a movable pulley for the lifting pendant, which pulley fits in a brass plate on deck (see page 94). This plate is apt to work loose under the strain of a heavy board, to avoid which the author devised the plan of fastening both lifting rod and pulley in their correct relative positions on a strip of hard wood. By this means the two are always in place, and the board may be lifted out or dropped in with a certainty that pulley and lifting rod are in their correct positions. The board is hung by two strips of sheet brass, the upper ends of both coming through the strip mentioned, being secured by a rivet. In use the board is hauled up, the pendant belayed on a cleat on the strip, and all may be lifted out together. With a board of 50lbs. it is sometimes all that a man can do to ship the lifting rod and pulley properly. but with this arrangement no special adjustment is necessary, the board and strip are dropped in place and all is ready. In trying this arrangement the idea of moving the board forward and aft suggested itself, and the following details were devised to accomplish it. The strip was provided with four small wheels, $a\ a\ a\ a$, sections cut from a $\frac{3}{4}$in. brass rod, with an $\frac{1}{4}$in. hole drilled in the center. These wheels work in slots cut in the slip A. The lifting pendant leads through the cheek block C, or through a double block if more power is needed, or the line may lead directly aft, without a purchase. A line G, is attached to the fore end of A, leading through a cheek block D, on the deck, thence aft to a cleat, as shown. A third line E, is fast to the after end of A, and leads through a hole in the coaming, as shown. In operation, the board is first dropped, then by casting off the line G the strip A, with board suspended from it, will run aft until the lifting pendant reaches the after end of case, when the board will be in the position shown by the dotted line. This will carry the center of the board aft about a foot at least, and will make a material difference in the balance of the hull and sails. The line E is used to draw the board aft if necessary, or a rubber spring may be attached to it, so as to act automatically. In hoisting, the board must first be drawn forward by the line G, after which it may be raised by F. There is this objection—the board cannot be raised if aground until it is hauled forward, but as the device is for sailing, usually in deep water, this is of little consequence. If the device is to be placed in a new boat, the case may extend as shown about a foot further aft, to the height of the boards, in which event the board may be raised some distance while in the after position, or may be easily cleared in case of grounding. The main use of the shifting board is to improve the balance of sail, allowing full or reefed sail to be carried at will, and giving just as much weather helm as may be at any time desirable. By its use the centerboard trunk can be placed well forward, and yet in racing the board itself may be readily adjusted to one of several positions, and may be thrown at least a foot further aft. Two points are necessary; the case must be wide enough to allow the board to move freely, and the after pendant must be made fast so far aft that the board will not rise at the fore end through the after end being the heavier.

The cheek blocks may be cut out of mahogany, with brass sheaves, or they may be cut or sawn out of $\frac{1}{16}$ in. sheet brass, filed up neatly and bent in a vise to the proper shape to fit the sheave. Sheaves of this kind are readily made by sawing with a hack saw pieces from the ends of brass rods of various sizes, holes are drilled in the centers, the pieces are held in a vise and the score or groove cut with a small round file. With a very few tools for working brass both blocks and sheaves with

AMATEUR CANOE BUILDING. 189

many other small fittings may be easily made by the amateur. The hoisting pulley B is shown on a large scale to illustrate the construction. The main part of the shell is made from a piece of sheet brass doubled over in the form shown, the upper part being a half circle. To each

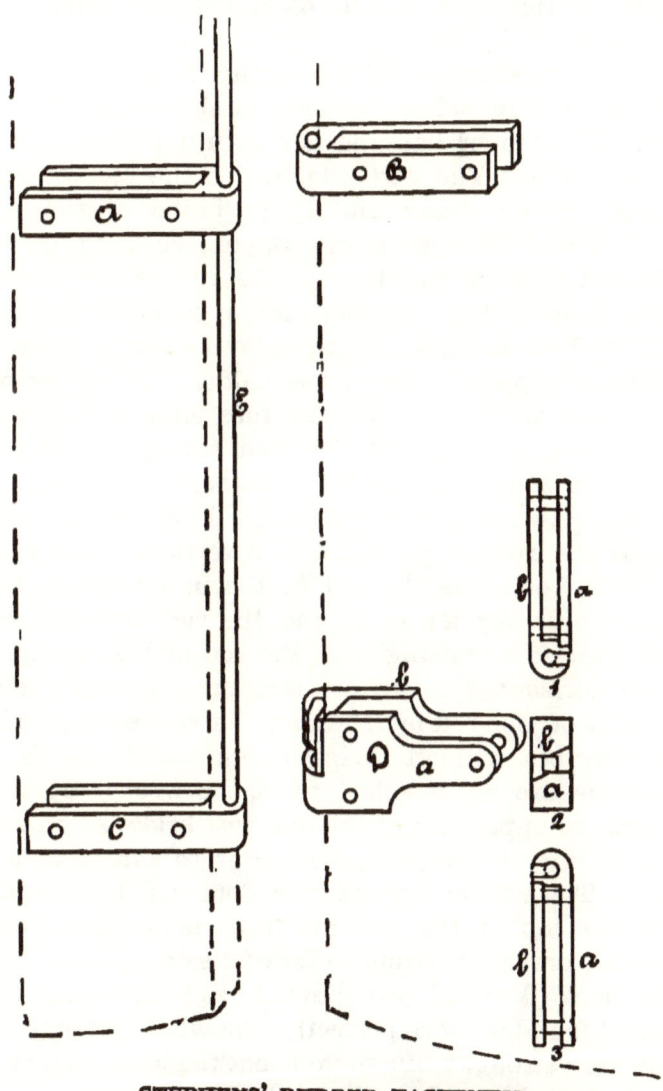

STEPHENS' RUDDER FASTENING.

side a strip of brass bent at right angles is riveted, a mortise is cut in the strip A, the brass case is set in until the side pieces rest on the strip, and then fastened by screws through these sides. The pulley is put in place and the pin on which it turns is run through holes drilled in the sides of the case, the ends then being riveted up.

The rudder shown is of mahogany, a cruising rudder. The lower side, below the keel, is sharpened to a fine edge. The rudder hanging is shown in the small drawing, page 189. The part attached to the boat or canoe consists of an upper and lower brace of the usual form, A and C, with a rod E, $\frac{1}{4}$in., running through them and screwed into C. On the rudder is a common brace, B, at the top. At the bottom is a split brace, D, made of two castings, a and b, both exactly alike, but fastened on opposite sides of the rudder. The upper sides of the pair are shown at 1, the fore ends at 2, and the lower sides at 3. It will be seen that by laying the rudder horizontally with the port side, D uppermost, the opening between a and b, Fig. 2, will admit the rod E. Now if the rudder be raised to a vertical position, the two hooks embrace the rod E, the upper brace, B, is dropped over the top of E, and the rudder is fast, only to be released by raising B off the rod and dropping the rudder horizontally. In practice the split brace can be put on or off the rod by inclining the rudder to an angle of 45 degrees, without laying it horizontal. With this gear there are no detached parts, the work may all be done at the upper part of the rod, just below A, and it is not necessary to grope under water to ship the lower pintle. The rudder can only be detached by raising B from the top of the rod, and the rudder lines, when attached, prevent it rising so far of itself.

The deck tiller and gear shown have been tried in practice and found to work perfectly, the whole arrangement being very strong, while there is not the least chance for lines to foul the rudder yoke or deck yoke. The former is in the shape of a wooden wheel, 6in. diameter and ⅜in. thick, turned in a lathe, with a groove ⅜in. deep around it, large enough to take the rudder lines of $\frac{3}{16}$in. copper wire rope. This wheel is fitted on its lower side with the usual dovetail plates, one being fastened to the deck and one to the wheel. The deck immediately beneath the wheel should be leveled, so that the wheel will bear on its entire lower surface. On the upper side of the wheel are two hasps, bent out of sheet brass and screwed fast to the wheel, in which the tiller ships. The tiller is reduced at the after end, where it fits into the after hasp, but on top of it is a spring, K, of flat spring brass, turned up at a right angle at the after end, as shown. This hook on the spring serves to hold the tiller in place, and further acts to raise the fore end of the tiller. By this means some spring is allowed in the latter, and in case of any weight being suddenly thrown on it, it will give until the hand touches the top of the hatch and will not break off. As the wheel has a bearing 6in. long, no matter what position it is in, there is no danger of twisting off the plate. The weak point of most deck tillers is the long pin and high block on which they are mounted to enable the yoke and tiller to clear the hatch, and with such a rig breakdowns are frequent. The present rig is both strong and compact, the long grotesquely curved arms that foul sheets and halliards are absent, and, if fitted closely to the deck, no lines can foul.

On the rudder head is a similar wheel, of the same size, but with the fore side cut away as shown, so as to allow it to be placed below the level of the deck. The angle allows ample play for the rudder. In order to insure perfect action the center of the wheel must coincide exactly with the center of the pin on which the rudder is hung, then the lines will be of the same length, no matter how far over the helm may be thrown. The lines for the foot gear, also wire rope, run round the rudder wheel, being fastened at one point only, so that they cannot slip on the wheel. The two ends are led forward and down through the deck, as shown. A brass ring is securely lashed to each line near where it leaves the wheel, and a strong hook on the end of each of the short lines from the deck wheel will hook into the ring. The short lines are each provided with turnbuckles, as shown, by which they may be tightened. When the deck tiller is not needed the turnbuckles are slacked up, the short lines unhooked, and the deck wheel may be removed, leaving only the lower plate set in the deck. The rudder wheel should be set as low as possible and yet allow the lines to clear the deck; there is no need to have it, as is often seen, far up in the air. If the two wheels are anywhere near the same height there will be no possible chance for the lines to run out of the grooves; in fact, if thrown out in any way they at once spring back. Of course there is nothing to catch the mizen sheet, as the lines will keep it from getting under the wheel. With a 6in. wheel there is power enough to turn a much larger rudder than is needed on a canoe; in fact, the wheel might be even smaller if desirable for any reason. One great advantage of a wheel over the ordinary arms is that the pull is always the same, no matter how far over the rudder may be.

The advantages of the old-fashioned leg-o'-mutton sail in the important points of light weight aloft and simplicity of rig, were so apparent that it is not surprising to find that while models and rigs were at first imported from England, the lug and gunter sails were soon discarded in America for the less complicated leg-o'-mutton, which, about ten years since, was the sail in general use here by canoeists. Both the lug and gunter required some care in rigging and more blocks and gear, but the plain triangular sail of the sharpie was easily made, after a fashion, and rigged to match, by any tyro. Up to 1878 this sail was used on almost all canoes in the United States, but as racing became more popular the lug was introduced and soon drove it out. The faults of the leg-o'-mutton sail were that the mast must be very long in order to obtain the area, and the sail was difficult to hoist and lower owing to the number of mast rings required, making it a slow and uncertain matter to reef or shake out. A mast of 14 to 15ft., such as was necessary for a 65ft. sail, was a very troublesome stick in a narrow boat, even if of bamboo. Naturally the idea of cutting this stick in half suggested itself, leading to the sliding gunter rig, but a mechanical difficulty was met that caused the gunter sail to be abandoned by all. It was found to be impossible to rig a gear of any sort that would slide on the lower mast and carry the topmast without either binding and jamming fast, or on the other hand, being so loose when hoisted as to allow the topmast to wobble to an unbearable degree. Brass slides of various forms were tried, as well as other devices, but besides the weight aloft, they never could be relied on to hoist and lower quickly, while they permitted a great amount of play in the topmast.

The sail here shown was designed last year by Mr. C. J. Stevens, New York C. C., for the canoe Tramp, and is also fitted to the canoe shown on Plate XXX. The sail plan on Plate XXIX. shows the rig adapted to the 15ft. × 30in. canoe on Plate XXVII. Curiously enough this new sail, a combination of the leg-o'-muttton and sliding gunter, was evolved directly from the balance lug. The first step was to sling the ordinary round-headed balance lug sail abaft the mast, of course retaining the peak, the yard being very much rounded, as shown in the sail plan of the *Forest and Stream* cruiser. This made a very good sail, but it was evident that the peak, falling more or less to leeward and out of the plane of the masts, was a decided disadvantage in so narrow a boat on the score of stability, whatever advantage it might possess to windward over a jib-headed sail. The next step then was to cut away entirely the angle between luff and head, the throat of the sail, substituting a moderate curve to the upper part.

The most serious difficulty was the hoisting and holding in place of the yard, now transformed into a sort of topmast; but after a number of trials a method was devised that is at once effective and ingenious. The object sought was to bind both mast and yard so firmly together that they became for the time a single stick, avoiding the play of the gunter, as well as the weak construction due to the short gunter brass with its direct strain on the m sthead. The details of the present plan are shown in the smaller drawing. On the mast two cheek blocks are securely lashed, leading fore and aft. On the yard are two similar blocks, also leading fore and aft. The halliard is double; a knot is first tied in the center, then the two ends are rove, one through each of the blocks on yard and then through corresponding block on mast, the bight of the line with the knot, c, being around the fore side of mast. It is evident that a pull on the two parts of the halliard will jam the spar firmly against the mast, practically making one spar of the two, as each braces the other. The size of each is so proportioned that the strength of the 2in. mast is retained all the way to the masthead, the yard growing larger as the mast grows smaller. The halliard was first used without the knot, c, but it was found that in hoisting the bight was held close to the mast, thus jamming at times. The knot was then tied in so that the halliard could not unreeve through either block; and now in hoisting the strain is only on one halliard until the yard is fairly in place, then both halliards are set taut and belayed. The canoeist takes both halliards in hand, leaving one with about 6in. slack, then hoists away, the yard rising easily, as the bight is entirely loose around the mast. When well up, a pull on both halliards, $a\ a$, sets all snug. The result is the same when set at the masthead for full sail, or lower for a reef. The boom and battens are fitted with jaws of the size and shape shown in the smaller cut, which represents the full size of the pattern, the casting being a little smaller when finished. The battens are round in section, and each is ferruled with brass, the jaw then being driven in. To hold the jaw to the mast a hollow brass curtain ring is used, lashed to the batten just on top of the jaw. This allows the boom or batten to be folded close up against the mast in stowing. The fourth batten may not be needed if the sail is properly cut; in any case it has no jaw, but is merely slipped into a pocket. The halliards, A A, lead from the masthead to blocks at deck on each side of mast, thence to a cleat near the well. They may be led through a double block near the stem, thus acting as a forestay, provided the mast is not too far forward; but the present practice is to place the mast from 9 to 12in. from stem, in some cases still closer. A tack line is used to hold the boom down, being led through a block at deck abaft the mast. The toppinglift is made fast to an eye at masthead, leading down each side of the sail, and spliced together just below the sheet block. A small jaw of brass is lashed to the boom, and the bight of the toppinglift is slipped into it, holding the boom at the proper height. By casting the lift off from this jaw and allowing it to swing forward the boom will drop on deck, and by leaving it in the jaw and hauling in, belaying on the small cleat or boom, the latter may be topped up as far as desirable. No reefing gear is shown, but any of the well-known varieties may be fitted as in a lug or Mohican sail, two reefs being sufficient.

The mizen may be rigged precisely as the mainsail, but with the small area now carried aft a sail of the same shape, but not fitted to lower, will answer perfectly. The luff is laced to the mast, one batten is fitted as shown, and a brail, in two parts, one on each side of the sail, and leading through small bullseyes lashed to the mast, the two parts spliced into one and belaying on the cleat at foot, serves to stow the sail snugly for running free or paddling. Though rather long, the mast and sail weigh but little, and may be readily unstepped and stowed on deck, a smaller mizen being substituted. The spirit mizen used on the Pearl will answer well for this latter, the spars being quite short. It is fitted as shown in the dotted lines, with one batten, there being rings on leech and head. A brail is rove through these rings in two parts, one and fast to boom and up the leech, thence through a bullseye on mast; the other end fast to top of batten, through rings on head, and through same bullseye. The sail may be reefed by hauling the latter part until the batten lies close to mast, or by a pull on both parts the sail is brailed up snugly against the mast. The spars may be short enough to allow the rig to be stowed inside the well.

The spinaker forms a most important feature of this rig, as the mizen is stowed when down wind and the spinaker set, the canoe running much steadier than under aftersail. The shape is shown in the drawing by the broken lines, the foot being greatly roached in the curve shown. The head and outer angle of sail are fitted with small swivels, in case of a turn in hoisting; the halliard leads through a block at masthead, one end leading through a block at deck, the other hooking into the sling of sheet block when not in use. If on the wrong side, it may be swung around in front of the mast before snapping to head of sail. The sheet or after guy is fast to the end, and a snap hook on the latter hooks into an eye on the end of boom. The tack leads through a screweye on deck just forward of the mast, the ends leading aft on each side of mast, so that either may be bent to the sail. The boom, of bamboo, is fitted with an eye at the outer end and a small jaw on the inner, the latter shipping in a brass stud in the deck just abaft the mast. The sail is snapped on to the halliard and hoisted, the tack being first hooked on; the outer angle is then hooked to the boom, the latter shipped against the stud on deck and swung forward, the sheet then being made fast and the tack trimmed. The sail should be of strong light linen that will dry quickly.

To complete the ordinary outfit for cruising and racing an intermediate mainsail of about two-thirds the size of largest sail is used, and sometimes a small spinaker.

The rig for a large canoe will include a racing mainsail of 90ft., a cruising mainsail of about 50ft., a racing mizen of 25ft., a cruising mizen of 15ft. and two spinakers of 60 and 45ft. The dimensions of spars and sails for such a rig are given in the following table, the spars being much lighter than any now in use on canoes, but they are all a little larger than those carried on the Tramp, a heavy Pearl, 14×33, for the past year; and if of good spruce and properly rigged, will be amply strong. The old Guenn carried a balance lug mainsail of 105ft and a mizen of 35ft., mast 15ft. above deck; but the present canoe, if built lightly and sailed with a moderate amount of ballast, should prove still faster under the rig shown. The weight aloft will be about one-half that of the old rig, consequently the boat can be held up with much less ballast and the crew will have far more control of her. The movement now is in the direction of smaller sails, and experience goes to show that a boat will be faster under a properly proportioned rig of moderate dimensions than under a heavy outfit of spars and canvas, that must be upheld by heavy ballast and at times with great difficulty:

DIMENSIONS OF SPARS AND SAILS—CANOE GUENN.

	Main.		Mizen.	
	Racing.	Cruising.	Racing.	Cruising.
Mast, from stem	11*	11*	12 00	12 00
deck to truck	11 00	8 00	9 06	5 02
Boom	10 06	8 04	5 07	5 00
Yard	10 06	8 04	....	....
Battens	9 01, 7 10, 6 0, 6 00	6 08, 5 02, 5 02	5 03	5 00
Spinaker boom	8 06	6 08	...	....
foot	10 00	7 11	5 03	4 08
luff	6 10	5 05	9 00	4 10
head	10 01	8 00	....	2 00
leech	17 06	13 10	9 04	4 04
Tack to peak	16 11	13 05	....	....
Clew to throat	10 11	8 08	....	....
Area, square feet	90 00	55 00	25 00	15 00
Spinaker area, sq. ft.	61 00	43 00	....	....
foot	10 00	8 00	...	....
luff	10 03	8 01	....	....
leech	13 04	10 06	...	....
round of foot	1 06	1 03	....	....
Battens apart	2 03, 2 04*	2 08	....	.

The 90ft. sail has three battens, 55ft. sail has two. Battens are spaced 1½in. further apart on leech than on luff.

Racing mast 2in. at deck and up to second batten, thence tapering to 1in. at head. Mizenmast 1⅜in. at deck, tapering to ¾in. at head.

Main boom 1⅛in. diameter for middle third of length, thence tapering to ¾in. at ends.

Yard 1in. diameter for about middle third, thence tapering to ½in. at ends. Battens round, ½in. at fore and ⅜in. at after ends, upper battens ⅜in. throughout. Mizenboom ⅞in. at middle, tapering to ½in. forward and ⅜in. aft., batten ⅜in. Spinaker boom, bamboo, about ¾in. at fore and ⅜in. at after ends. The spars for cruising rig will be a little smaller throughout. The mizenmast will fit forward tube, but will be reduced in size from deck up.

A method of leading the reeflines, devised by Mr. O. F. Coe, of Jersey City, is shown in the following sketch.

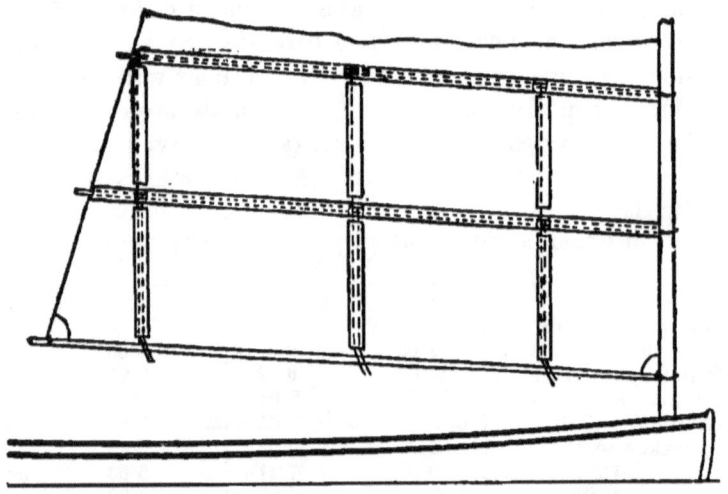

Vertical pockets are sewn to the sail through which the lines are run, thus lessening the danger of fouling. Mr. W. Baden-Powell has used the same idea for some time, but with several small brass rings sewn in the pockets to keep them extended and so allow the lines to run freely.

The drop rudder is now in general use for cruising as well as racing and is made after several patterns by the different builders. One of the best is that shown in

Plate XXIX., made by Chas. Piepenbrink, of Albany, New York. The stock is a brass tube, A, ⅜in. external diameter, into which two side pieces of $\frac{1}{16}$in. sheet brass, C, C, are inserted and brazed, sufficient space being left for the blade B to work freely between them. The blade, of $\frac{1}{16}$in. hard brass, turns on the flat-headed bolt, F, and is held more firmly by the lugs E E, riveted to each side and projecting over the side plates. The rudder yoke, D, is of cast brass, very neatly proportioned in its design, with eyes at each end for the rudder lines. It is held on the stock, A, by a set screw, I, passing through A, and in the upper end of the latter is an eye through which the lifting line, K, is rove. The rudder is hung by two braces, the lower one, H, forming a scagband and at the same time having a hole for the ⅜in. pin in the lower end of A. The upper brace, G, is bored out to ⅜in. diameter inside, with a slot at the back to allow the side pieces to enter. On the side of A is a small stud, I, which also passes through the slot, and when the rudder is in position prevents it from rising. Hard brass only should be used for the side plates and blade, as great stiffness is necessary.

On plate XXIX.a are shown the latest fittings used on American canoes. The upper cleat, invented by Mr. Paul Butler, is very handy for main sheet, a turn being taken under the hooked end. A somewhat smaller cleat, devised by Mr. E. H. Barney, is also shown. The cleat board introduced from the English canoes, is now generally used. It is a piece of mahogany 3 or 4in. wide and long enough to extend across the cockpit, to which it is secured by the hook screws shown, which allow it to be shifted to any point. In this board are belaying pins, as shown, or cleats are screwed to it, and sometimes a traveler of ¼in. wire is added. In removing the sail all lines are left on their respective cleats, the board being detached and made up with the sail. The tiller shown was fitted to the Blanche by Mr. Butler, the ends extending across the boat so that one is within easy reach when the crew is leaning out to windward. The ordinary tiller may also be added, as shown by the dotted lines. Two varieties of lifting handles are also shown, the one devised by Mr. Barney for his 1887 canoe, Lacowsic, being of brass, set into the stem and stern of the canoe, which are cut away to receive them. The other handle, a piece of stout wire with a short length of rubber tube for the hand to grasp, is used on most of the Mohican canoes, being permanently attached one to each end. The mast and jaw shown are those of Pecowsic and Lacowsic, the tube is but 5in. deep and the lower end of mast is fitted with a long taper ferrule to fit it very neatly. The jaw is of brass, with a socket for the boom end, and is hung by a pin through the mast, allowing the boom to be folded close against the latter. The mast turns in the socket, the boom being immovably fixed to it. The mast lacing shown is described with the sail plan of Notus, Plate XXVI.

PLATE XXX.—16×29 RACING CANOE. DESIGNED BY W. P. STEPHENS.

This canoe was designed early in 1888 by the author as a racing craft, to be sailed without ballast; the displacement being limited to not over 275lbs. The aim has been to preserve a good area of load water plane and breadth, without too much displacement on the one hand or the sharp V sections of some "no ballast" canoes on the other, and to make a canoe that should be at the same time fast and yet fairly comfortable to sail. The design could easily be adapted to a larger canoe, say 16×30, to carry moderate ballast, by widening after planking, and building up the sheer line an inch or a little over, making at the same time the displacement greater by 100lbs. and the draft nearly an inch more. The canoe has been completed and will be raced during the season of 1888. She will carry the sails shown in Plate XXIX., with a very light plate board and brass drop rudder. As she is designed solely for sailing the well will not be as shown, but simply a circle of 18in. diameter, closed by a watertight canvas bag made fast around the coaming, so that in the event of a complete capsize no water can get below. The deadwood at the ends has been cut away far more than is common, but the displacement is somewhat reduced thereby, more being allowed in the middle of the boat; the immersed surface is much reduced; and easy turning power insured, the full length on waterline being at the same time retained.

TABLE OF OFFSETS, 16×30 RACING CANOE.

Stations	Heights			Half Breadths						
	Deck	Rab't	Keel	Deck	10in.	8in.	6in.	4in.	2in.	Rabbet
0..	1	3^4	.....	0	0^1	0^1	0^1	0^1	.....	0^4
1..	1	2^2	2^7	2^3	3^3	2^7	2^5	2^1	1^2	0^4
2..	1	1^4	1^4	1^2	6^2	5^5	5	4^1	2^7	1^1	0^4
3..	1	0^5	0^5	0^6	8^4	7^1	7^2	6^2	4^4	2^8	1^1
4..		11^7	0^2	.....	10^2	10	9^2	8^2	6^4	4^2	1^4
5..		11^2	.....	.....	11^7	11^5	11^1	10^1	8^4	5^4	1^4
6..		10^7	.....	.....	1 1^1	1 1	1 0^4	11^6	10^2	7^2	1^4
7..		10^4	.....	.....	1 2	1 1^7	1 1^4	1 0^7	11^5	8^5	1^4
8..		10^2	.....	.....	1 2^3	1 2^3	1 2^2	1 1^6	1 0^5	9^4	1^4
9..		10^1	.....	.....	1 2^4	1 2^4	1 2^3	1 2^1	1 0^7	9^6	1^4
10..		10	0^1	.....	1 2	1 2	1 2^2	1 1^7	1 0^6	9^2	1^4
11..		10^1	0^3	0^1	1 1^7	1 1^7	1 1^5	1 1^1	11^6	7^6	1^2
12..		10^3	0^6	0^4	1 0^6	1 0^4	1 0^2	11^3	9^4	5^2	1^1
13..		10^4	1^2	1	11	10^6	9^7	8^6	6^5	2^1	0^7
14..		11^2	2	1^4	8^1	9^4	6^4	5^2	3^4	0^5	0^5
15..	1	0^1	2^7	2^2	4^4	3^4	3	2^1	1^1	.....	0^4
16..	1	1	.....	3	0^1	0^1	0^1	0^1	.....	0^4

The dimensions are:

```
Length...........................................16ft.
Beam, extreme.................................29in.
      l.w.l...................................25¾in.
Draft............................................4½in.
Freeboard......................................6in.
Sheer, bow.....................................5½in.
      stern.......................................3in.
Displacement...............................256.75lbs.
Area midship section......................48 sq. ft.
Displacement per in. immersion.........109lbs.
          at 5in. draft, about............370lbs.
Area lateral plane.........................4.37 sq. ft.
      centerboard............................2.17 sq. ft.
      Total.....................................6.54 sq. ft.
L.W. plane...................................19.60 sq. ft.
C. B. from stem..................................8ft.
C. L. R. from stem........................7ft. 10¾in.
      inc. board..............................7ft. 5in.
```

PLATES XXXI. AND XXXII.—18×36 CANOE "IONE."

The canoe Ione was designed in 1887 by Mr. E. T. Birdsall, of New York, and built by Bradley, of Watertown, N. Y.

She is 18ft. long by 3ft. wide and is smooth built, of ⅜in. plank; keel of white oak 1¼in. thick; stem and sternpost of hackmatack; planking of cedar; deck, mahogany; coaming, walnut, flared; Radix board of largest size, which is rather small for a boat of this length. She carries about 125ft. of sail and 100lbs. of lead cast to fit close to the garboards, no shifting ballast. With this lead in and sail set, one can sit outside of the coaming on the deck to leeward and not get wet, the sails and spars weighing 50lbs., and the usual crew over 150lbs. In a beam wind in smooth water with full sail and two persons weighing together 340lbs. hanging out to windward she has beaten catboats of equal waterline length. When driven in heavy weather with the above load to windward and reefed she is quite wet, as she goes through the waves when they are short and choppy, and the crests thus cut up come aboard.

In common with all narrow and shoal boats she rolls when going to leeward in a seaway. The under-water body conforms to the wave form curve of areas, other recognized principles of design as set forth by Dixon Kemp being adapted to this special case as far as possible.

TABLE OF OFFSETS—CANOE IONE.

Stations.	Depths.			Half Breadths.							
	Deck	Rab-bet.	Keel	Deck	12	10	8	6	4	2	Keel.
0..	2			0^1	fore side of stem						
					0^1	0^1	0^1	0^1	0^1	0^1	
1..	1 10^2	2	1^4	3^5	2^7	2^4	2	1^4	0^7	0^4	0^4
2..	1 8^6	1^5	1^2	7^1	5^5	5	4^2	3^2	2^1	0^5	0^5
3..	1 7^2	1^4	1	0^1	8^3	7^5	6^6	5^4	3^6	1^4	0^7
4..	1 5^5	1	0^5	1 0^2	11	10^2	9^2	7^6	5^5	2^5	1
5..	1 4^3	1	0^4	1 2^2	1 1^1	1 0^3	11^4	10^1	8	4^3	1^5
6..	1 3^5	0^6	0^2	1 3^5	1 3^1	1 2^4	1 1^5	1 0^3	10	6	2^1
7..	1 2^4	0^5	0^1	1 4^5	1 4^3	1 3^7	1 3^2	1 2	1	7^7	2^2
8..	1 1^7	0^4		1 5^3	1 5^1	1 4^6	1 4^2	1 3^2	1 1^4	10	2^3
9..	1 1^4	0^4		1 5^6	1 5^5	1 5^3	1 5	1 4^1	1 2^6	11^1	2^4
10..	1 1^3	0^4		1 6	1 5^7	1 5^5	1 5^3	1 4^5	1 3^2	11^5	2^4
11..	1 1^5	0^5	0^1	1 5^6	1 5^6	1 5^3	1 5	1 4^2	1 2^5	11^1	2^2
12..	1 2	0^7	0^3	1 5^2	1 5^1	1 4^7	1 4^2	1 3^2	1 1^4	9^2	2
13..	1 2^7	1^1	0^5	1 4^3	1 4^1	1 3^6	1 3	1 1^5	11	5^6	1^4
14..	1 3^7	1^2	0^6	1 2^5	1 2^3	1 1^7	1 1	11^3	8^2	3^2	1^2
15..	1 5	1^5	1^1	1 0^4	11^5	11	9^7	7^6	5	1^5	0^7
16..	1 6^5	1^7	1^3	9^3	7^5	6^6	5^4	4	2^4	0^6	0^5
17..	1 7^2	4^1	1^5	5^4	3^2	2^5	2	1^2	0^1	0^2	0^2
18..	1 10			after side of sternpost							
				2	0^1	0^1	0^1	0^1	0^1	0^1	

During the season of 1887 she was sailed with the ballast and sails shown in Plate XXXII., with either one or two as crew, but a jib of 30ft. will be added, cutting the present mainsail down the line of the mast, thus making it a gaff sail all abaft the mast, the bowsprit to be 4ft. outboard and the jib to trim aft of the mast and be capable of being set and taken in from the cockpit. When sailing alone, in addition four 25lb. pigs of lead cast in the form of a truncated pyramid and covered with canvas and roped, will be carried to be shifted to windward.

Ione has no watertight compartments, but her owner proposes to fit them in her. The sails are of Polhemus twill, about 4oz. A 6lb. Chester anchor and 25 fathoms of 12-thread manilla has held on in 15 fathoms of water with a good jump on, a lee-going tide and two 15×30 canoes fastened to mizenmast. All the sailing of the above boat has been done in Newark and New York bays and around Sandy Hook and Staten Island. The yards are egg-shaped and the booms are square to facilitate the reefing gear.

CANOE YAWLS.

The success of the earlier canoes called the attention of boating men generally to the many good points of this type of boat, with the result that a number of large craft have been built much on the lines of the ordinary sailing canoe. All of the earlier boats were yawl rigged, some like a canoe and others with a jib, and hence the name "canoe yawl" was given to distinguish them from the small canoe. Within the past three years the number of these boats has greatly increased in England, while they are also becoming better known and liked in America, and some of them make very fine cruising craft, being far more able and powerful than the canoe. They are built with centerboard or keel, generally the latter, and are rigged with the main and mizen, like the canoe, or as cutters, sloops and yawls, the latter being perhaps the best for single-hand cruising. Some of them, such as the Cassy, the Water Rat and the Viper, have made their reputations as cruisers by several seasons of constant work in open waters. They are well fitted for bays and arms of the sea where the canoe cannot safely and comfortably be used; their shape, that of the whale boat and surf boat, is one of the best for a sea boat, and they are less costly to build than the small counter-sterned yacht, while superior to the square-sterned boat.

PLATE XXXII.—"ANNIE," CENTERBOARD CANOE YAWL.

This boat was designed to have a light draft and to be light enough to house easily, so a fixed keel and ballast were dispensed with. Her leading dimensions are: Length, 18ft.; beam, 5ft. The drawings show a slightly smaller boat, but a scale was used in building which brought the beam up to 5ft. Annie was built at Oswego, N. Y., for Mr. Geo. N. Burt, by Joseph Henley, who made the model from the owner's instructions. She has been used on Lake Ontario with great success, proving fast as well as safe and comfortable for pleasure sailing. As the hull is light it can readily be hauled in or out of the house by one man, quite a consideration in some localities. Annie is planked with $\frac{7}{16}$in. cedar and white pine in alternate streaks, the timbers being $\frac{3}{8} \times \frac{1}{2}$in., spaced 4in. The deck is of cedar, on chestnut carlins $1 \times 1\frac{1}{4}$in., spaced 6in. The cockpit is 7ft. 2in. long and 3ft. 5in. wide, with a 3in. coaming of butternut. The centerboard trunk is 3ft. long and the board is of boiler plate, 26lbs. The total weight of hull is 300lbs. The ballast consists of six bricks of lead, 25lbs. each, stowed in the space abreast the trunk, besides which two bags of sand, 50lbs. each, are carried in the well. The lead bricks are covered with canvas and have rope handles, so they are quickly carried in or out. The spaces in each end are filled with air tanks, one being placed also on each side of the well as shown. No oars are used, a paddle being carried for calm weather, but the boat is expected to sail whenever there is any wind. She is rigged with a boom and gaff mainsail and a sprit mizen. The mainmast is 17ft., heel to head, and 3in. in diameter; mainboom 12ft., gaff 5ft. 4in., mizenmast 11ft., and 2¼in. in diameter, boom 5ft. The hoist of mainsail is 12ft. 4in., and of mizen 8ft. The main gaff has peak and throat halliards, the former with double block on mast and single on gaff. Both halliards lead through fairleaders on deck to the after end of trunk, where they belay. The rudder is fitted with long steering lines. There are no fixed thwarts, but movable seats are used. A spinaker is carried on the mainmast, the boom being jointed for stowage. She has been through some bad weather on Lake Ontario, proving herself a fine rough-water boat, riding lightly and going well to windward in rough water. In ordinary sailing she is very fast, and with two or three persons aboard carries her sail easily.

Plate XXXIII.—"Cassy."

Length	14ft.
Beam	3ft. 4in.
Depth midships	1ft. 4in.
Sheer, bow	11½in.
stern	5in.
Bow to after side of tabernacle	3ft.
fore end of trunk	4ft.
after end of trunk	7ft.
after end of well	11ft. 6in.
rowlocks	9ft. 6in.
Area, mainsail—racing	120 sq. ft.
mainsail—cruising	60–70 sq. ft.
mizen	15 sq. ft.
Length of tabernacle	18in.
oars	8ft.
Width of rudder	1ft. 6in.

The canoe yawl Cassy was designed and built by Mr. G. F. Holmes, for use on the Humber River. She is fitted with the tabernacle and centerboard devised by Mr. Tredwen, the latter of 70lbs. being all the ballast used with cruising rig, but sandbags are carried in racing, about 100lbs. being used. The forward thwart can be placed 2in. below the gunwale for rowing, or about 6in. above the bottom for sailing. The rig includes two balance lugs as in a canoe, with a deck tiller.

The smaller cut is described on page 23.

Plate XXXIV.—"Vital Spark."

The Vital Spark is of canoe model, 18ft. long, 5ft. beam, 2ft. 2in. draft. She is carvel built, with ¾in. planking, keel sided 3½ at middle, 1¾ at ends, with 4½cwt. of lead underneath, and an equal amount of lead inside.

The sail plan is that of a similar boat, the Viper, whose sheer plan and rig are shown in the drawing; she is 20ft. long, beam 5ft. 5in., depth to gunwale amidships, 2ft. 6in. Deck has a crown of 5in., and is of light wood covered with canvas. Her keel has 19cwt., 2qrs., 19lbs. of lead, with 2cwt., 1qr., 18lbs. inside, and an iron keelson of 75lbs. The depth of keel is 1ft. 9in., and the total depth 3ft. 4in.

The rig is a convenient one for small boats, as jib and mizen may be used together in strong winds, the mainsail being stowed. The Viper, as shown, carries a staysail as well as a jib, and a small gaff trysail.

PLATE XXXV.—SAIL PLAN OF CANOE YAWL.

The term "yawl" applied to a cutter-rigged boat is an anomaly, but the type of boat in question is now commonly known as the "canoe yawl," from its derivation directly from the canoe and the fact that it is almost invariably yawl rigged. The boat shown in Plate XXXV. was built from the lines of the Vital Spark, Plate XXXIV., but was rigged as a cutter. Her dimensions are as follows:

Length on deck	18ft. 4in.
Beam	5ft.
Draft	2ft. 2in.
Freeboard	1ft.
Cockpit	7ft. 6in.×4ft.
Lead keel	850 pounds.
Ballast inside, iron	250 pounds.
Planking	¾in.
Mast, from fore side of stem	6ft. 10in.
Mast, deck to hounds	14ft. 3in.
Mast, deck to truck	18ft. 9in.
Mast, diameter at deck	4in.
Bowsprit, outboard	6ft.
Bowsprit, diameter at stem	3in.
Mainboom	15ft.
Mainboom, diameter	2½in.
Gaff (oval, 2¼×1½in)	9ft. 6in.
Center of lateral resistance aft center of loadline	10in.
Center of effort forward of center of loadline	5in.
Center of effort above loadline	6ft. 4in.

SAIL AREA.

	Foot.	Luff.	Leech.	Head.	Area.
Mainsail	13.9	11.0	17.8	8.9	150 sq.ft.
Staysail	8.2	13 0	11.3	...	48 sq.ft.
Jib	8.0	15.8	11.0	...	45 sq.ft.

Total sail area	233 sq.ft.
Area of reefed mainsail	80 sq.ft.

With the above amount of ballast the draft is a little less than 26in., but in cruising the crew and stores would bring her to her loadline. The center of effort of reefed mainsail and whole staysail is shown at C E 2, and of the two headsails at C E 3. Many will object to the double rig, but in practice it is found to work excellently, being very easily handled. The three small sails are easily set by a boy, and the headsail sheets, leading to the rail as shown, may be reached from the tiller. In tacking they are readily got down with one hand without leaving the stick. The jib is set flying, the outhaul being an endless line, with a snaphook spliced in. The hook is snapped to the jib tack, the sail partly hoisted and hauled out. When not in use it is stowed in a bag instead of being furled on the bowsprit. No jibstay being needed, the bowsprit is fitted with a tackle on the bobstay and is easily housed entirely, which is sometimes a great convenience in running into odd places as such small boats constantly do. The fittings are very simple, a gammon iron bolted to port side of stem head, a sampson post of 2x6in. oak plank, with a 3¼in. hole bored through for the heel of the round bowsprit, a fid of ½in. round iron, and two small iron blocks for the bobstay tackle, one hooking into a wire rope bobstay.

In some cases a tabernacle and lowering mast are desirable, and with a forestay both are easily fitted. The tabernacle is made of two pieces, B B, of oak 1¼x4 inches, stepped in the keel, D, and coming to the coaming I I. The mast is stepped in the block C under the floor K, and is held by the forestay and two shrouds, all fitted with turnbuckles. A bar F of 1¼x¼in. iron is bolted to the tabernacle's sides, one bolt G being fitted with a thumb nut, while the bar is slotted on the starboard side to slip over the neck of the bolt, turning on the port bolt. When G is loosened the bar may be turned over out of the way and the mast lowered. To avoid cutting away the floor for a distance aft of the mast, a block of oak, E, is bolted to the heel of the latter, on the after side. When the mast is lowered the block turns on the edge L, lifting the mast out of the step as it falls aft. In lowering, the halliards are stopped to the mast out of the way, the jib halliard is carried forward and hooked to stem head, the bar F is swung back and the mast is lowered by the jib halliard. The shrouds and also the parrel on the gaff must both be slackened. One man can readily lower and hoist the mast for bridges, etc.

The leads of the various lines are as follows: Throat halliards to cleat d on starboard side, peak to cleat b on same side so that both can be reached at the same time; staysail halliards on cleat c, jib on cleat a, toppinglift on cleat e on mast, staysail downhaul knotted in hole in coaming at f. The mainsail is thus set from the starboard and the headsails from the port side of the boom, and the downhaul is handy to the staysail halliard. All are easily reached by leaving the tiller for a moment, and one man can manage all lines. The boat has air tanks in each end, a large cuddy forward and seats in the cockpit. For cruising the seats would fold out, making a bed for two or even three (4x7ft.), while a tent would be pitched over the boom. The yawl rig would answer well for such a boat, but the present one has proved very satisfactory for singlehanded sailing and cruising.

The following descriptions of similar boats are given by correspondents of the London *Field*, in answer to inquiries:

One writer says: "I have just launched a canoe yawl, length 18ft. by 5ft. 8in. beam, and a draft of 2ft. aft, and 1ft. 3in. forward. She has at present 9cwt. of lead and iron ballast inside, but requires 4cwt. or 5cwt. more. She is fitted with a well 7ft. 6in. in length, the fore end being 7ft. 6in. from the fore side of the stem. She is rigged with a standing lug mainsail, hoisted with a single halliard, and the tack purchased down with a gun tackle; the clew is hauled out with a traveler on the boom; which is fitted to the mast with a gooseneck; the mast is stepped 2ft. 6in. aft of the outside of the stem; the mizzenmast is stepped 1ft. inboard from the stern, the sail being a leg o' mutton. Height of mainmast above deck, 19ft.; height of mizzenmast above deck, 10ft.; length of head of lug, 14ft.; length of luff of lug, 9ft.; length of leach of lug, 23ft; length of foot of lug, 13ft. 6in.; length of luff of mizzen, 8ft 6in.;

length of leach of mizzen, 8ft. 6in.; length of foot of mizzen, 6ft. On the trial trip she handled very well under sail; with the tiller amidship, she nearly steered herself on a wind. In placing the well aft I secured room for a comfortable little cuddy under the foredeck, with a headroom of 32in.; and with only 3ft of deck aft of the well, I do not require a deck yoke steering gear, as used on the Mersey canoe yawls, but have an ordinary iron tiller, with a crook in it to pass the mizzen. I think 'Pansy' could not find a much handier rig for this class of boat. I may mention that I have had the above canoe yawl built for use on the Humber."

Another adds: "In reply to 'Pansy,' permit me to say that I have sailed single-handed for some years a Mersey canoe with a center plate, nearly the same size as 'Pansy's,' under a standing lug and mizzen, and a handier, safer and more seaworthy little craft I could not desire. She was built here very faithfully and cheaply. I have, however, found that the sail originally given her was too much for real sea work, although considerably less in area than Mersey canoes are designed theoretically to carry. My ballast was 370lbs. lead inside, and the iron plate weighed 110lbs. The sail I tried to carry at times was a lug with a boom 10ft. on head and foot; luff, 5ft.; leach, 14ft.; jibheaded mizzen, foot, 5ft.; luff, 6ft. 6in.; leach, 7ft; height of mainmast, step to truck, 12ft. But seldom indeed could I give her this sail when single-handed, so I reduced the inside ballast to about 112lbs., the mainsail to 8ft. on head and foot for light winds, and had another lug 6ft. 6in. on head and foot, with 4ft. luff and a reef in the mizzen for every-day work; under the latter sails the boat was, all round, more useful and infinitely drier in a sea way. Guided by rough experience, I advocate for single-handed small boat work, the lug and mizzen sail plan, with a shift of main lugs (the lug set by Dixon Kemp's plan of peak and throat halliards, which is admirable indeed), in preference to lug, jib and mizzen. Simplicity is the true motto for single-handed small boat work at sea; and if a boat is equally handy without the head sail, why should gear be complicated with head sheets and halliards?"

Plate XXXVI.—Mersey Canoes.

The Mersey canoes or canoe yawls, have grown out of the small canoes, and are used like them for general cruising, but on more open waters. The dimensions are: Length 17ft., beam 4ft. 6in., depth 2ft. Oars are used, as the beam is too great to admit of paddling. The deck and well is similar to a canoe. Lead ballast is stored under the floors. The rig consists of two lugs, main and mizzen, the dimensions being:

	Racing mainsail. Ft. In.	Cruising mainsail. Ft. In.	Mizzen. Ft. In.
Foot	10 00	6 06	4 06
Head	10 00	7 06	2 03
Luff	5 00	2 06	2 04
Leach	14 00	10 00	6 00
Tack to peak	14 08	9 00	5 00
Clew to throat	10 09	7 00	4 09

As there is no centerboard the interior of the well is entirely unobstructed, and there is room for three persons, though on a cruise two, with the necessary stores and baggage, would be enough. Beds for two might easily be made up on the wide, flat floor, a tent being pitched over the well, while the seats may be removed entirely at night. Under the fore and after decks is ample room for storage of all stores. The steering is done with a deck tiller, as in a canoe.

In building such a boat, the stem, sternpost and keel would be of oak—or the former of hackmatack—sided 1¼in.; keelson of oak, 3x¼in.; plank of cedar, ⅙ or ⅜in. lapstreak; gunwale of oak or mahogany; deck of ⅜in. pine, covered with 6 to 8oz. drill laid in paint; coamings of oak, ⅜in. thick. The ribs would be ⅜x⅜, spaced 9in., with floors at every alternate frame.

The sails are rigged as "standing lugs," or a yawl rig similar to the Viper may be carried. They will be of 6oz. drill, double bighted; rigging of "small 6-thread" manilla; blocks of wood, iron or brass.

The dimensions of a similar canoe are given in "Cruises in Small Yachts and Big Canoes," by Mr. H. F. Speed, as follows: 16ft. long, 4ft. 1¼in. beam, 20in. deep amidships, with

6¼in. of keel, containing 3cwt. of lead. Inside she carried 1cwt. 10lbs. of lead. The sail area was 180ft. mainsail and mizen, lugs, with jib, the dimensions of spars being:

```
Main mast.............................................13ft. 1in.
    boom for lug sail...........................10ft. 4in.
    yard for lug sail.............................12ft. 6in.
    boom for gaff mainsail................... 8ft. 5in.
    gaff for gaff mainsail....................... 8ft. 6in.
Mizen mast.......................................... 8ft.
    boom............................................... 6ft. 4in.
    yard................................................. 7ft. 4in.
    boomkin, outboard......................... 2ft. 6in.
Bowsprit, outboard................................ 5ft. 9in.
Spinaker boom......................................10ft. 6in.
Tonnage, "one ton and an awful fraction."
```

Her well was 5ft. 6in. long and 2ft. 6in. wide, with a locker aft for stores, open lockers along the side, and two shifting thwarts, steering with a half yoke on the rudder, and a rod hinged thereto, the motion, of course, being fore and aft. The well was covered completely by a tent.

Plate XXXVII.—"Iris" Canoe Yawl.

This boat was built in 1887 by J. A. Akester, of Hornsea, near Hull, Eng., and is now owned by Mr. Holmes, of the Cassy. The hull is carvel built. The mast is fitted with a tabernacle for lowering, the sail plan being shown in plate. The inside ballast is in four blocks, two being generally carried, while the lead keel weighs 450lbs. A centerboard could readily be fitted to work entirely beneath the floor, and would be a great aid to the boat in windward work. The tiller is of iron, and curved as shown so as to work about the mizen mast. The dimensions are as follows:

```
Length over all.....................................18ft.
    l.w.l. .............................................17ft. 4 in.
Beam, extreme.................................... 5ft. 1 in.
    l.w.l............................................... 4ft. 7 in.
Draft, extreme..................................... 1ft. 4½in.
Least freeboard................................... 1ft. 1 in.
```

Sheer, bow.................................. 10^{4}in.
stern...................................... 7 in.
Ballast, keel, lead.............................450lbs.
inside, lead ....225lbs.
Mainmast, from stem....................... ...... 2ft. 3^{4}in.
deck to truck........................15ft. 3 in.
Mizenmast, from stem........................17ft. 3^{4}in.
deck to truck..... .... 7ft.
Mizen boomkin................................ 2ft. 8 in.
Main boom....................................15ft.
yard ....15ft.
Mizen boom............ .... 6ft.
batten.. ... . 6ft. 6 in.
Mainsail, area.... ....168 sq. ft.
Mizen, area .... 25 sq. ft.
Total..193 sq. ft.

TABLE OF OFFSETS—CANOE YAWL IRIS.

Stations.	Heights.			Half Breadths.						
	Deck	Rab't	Keel.	Deck	No. 1.	No. 2.	L WL.	No. 4.	No. 5.	Rabbet
0..	3 4			0^2						1
2..	3 0^2	11^6	8	1 0^7	8^4	7	5^2	3^2		1^9
4..	2 9	9^4	5^2	1 9^1	1 6	1 4	1 1^2	9^4	4^9	2^5
6..	2 7^1	7^7	3	2 2^2	2 0^5	1 11^1	1 8^6	1 5	11^3	3^1
8..	2 6^2	7	1^4	2 5^3	2 4^4	2 3^4	2 2	1 11^2	1 0^4	3^2
10..	2 5^6	7	0^4	2 6^2	2 5^6	2 4^7	2 3^4	2 1^4·1	8^4	3^2
12..	2 6	6^6		2 5^5	2 4^7	2 4	2 2^4	1 11^6	1 7^2	3^1
14..	2 7^4	6^6	0^2	2 2^2	2 0^6	1 11^5	1 10	1 7^1	1 2^1	3
16..	2 9^7	6^6	1^1	1 5^5	1 3^3	1 1^7	1	9^4	6^6	2
18..	3 0^4		2	0^3						1

SNEAKBOXES AND CRUISING BOATS.

Plate XXXVIII.—Barnegat Sneakbox.

This curious boat is used on Barnegat Bay, on the New Jersey coast, for duck shooting and sailing. Being low on the water, it is easily converted into a blind, by covering with brush, and its flat, spoon-shaped bottom allows it to be drawn up easily on the mud or sand. The usual size is 12x4ft., and the rig is a small spritsail. Most of the Barnegat boats are fitted with a dagger centerboard, sliding in a small trunk from which it may be drawn entirely. The boat shown is used only for sailing, and is fitted with a board of the usual form, hung on a bolt. The rig is also different. The dimensions of the boat are: Length over all, 16ft.; beam, 4t. 11in.; depth amidships, 12in.; draft, 8in.; keel, of oak, 5in. wide; frames 1¼in., sided, 1in. moulded, spaced 13in.; planking (carvel build), $\frac{9}{16}$in.; round of deck, 8in.; deck planking, $\frac{3}{8}$in.; coaming 2in. high at sides; width of rudder, 24in.; mast at deck, 3in.

The sail is a balance lug, hung from a single halliard. The yard and boom are each held in to the mast by parrels. There is no tack to hold the boom down, as is usual in these sails, but a line is made fast to the free end of boom, leading to the deck near the mast, where it is belayed, thus preventing the sail from running forward, and answering the purpose of a tackline. The dimensions of the sail are: Foot, 15ft. 6in.; head, 9ft. 8in.; luff, 9ft. 7in.; leach, 20ft.; clew to throat, 16ft. 10in. Area, 160 sq. ft.

In anchoring the boat the cable must be rigged so as to be reached from the well, as in the smaller sizes a man cannot walk out on deck to the bow.

PLATE XXXIX.—THE BARNEGAT CRUISER.

The prominence given to cruising of late years by the increased growth and importance of canoeing has brought many into the ranks who do not care for so small a boat as the canoe, but who wish a strong, roomy and serviceable boat, either for cruising or for general sailing, alone or with one or two friends. The canoe is really a boat for one person, and must be capable of being paddled, sailed and handled on shore by one; but where these conditions do not prevail a different type of boat may often be used to advantage. In places where the boat must lie afloat and a fine canoe would be damaged; in open waters where there is no occasion to haul up the boat, and where transit by rail is not an object; in shooting trips and other cruises where several persons, and perhaps a dog, must be carried, the boat shown in Plate XXXIX. will answer admirably. This boat, named by its designer the "Barnegat Cruiser," is an adaptation of the common sneakbox, found all along the New Jersey coast (see Plate XXXVIII., p. 215) to the wants of cruisers. Mr. N. H. Bishop, so widely known as a canoeist, cruiser, traveler and an able writer in behalf of cruising, has for some time past made a special study of the sneakbox at his home at Toms River, N. J., the home of the craft. He has built and tried boxes of all kinds, has experimented largely with sails, and has expended considerable study on the details of fittings. The boat shown in Plate XXXIX. is a 14-ft. cruiser of the new model, a number of which are now building at Toms River under Mr. Bishop's personal supervision, for members of the American Single-hand Cruising Club. The lines shown were taken from "Seneca's" boat, described in the *Forest and Stream.* The dimensions are:

Length	14ft.	0in.
Beam	4ft.	6in.
Depth at gunwale	1ft.	1in.
Sheer, bow		8½in.
Sheer, stern		4in.
Draft, loaded		6in.
Freeboard		7in.
Crown of deck		8in.

AMATEUR CANOE BUILDING. 217

Fore side of stem to—
Mast tube......................................	2ft.	0½in.
Trunk, fore end............................	3ft.	1in.
Trunk, after end............................	6ft.	3in.
Well, fore end...............................	5ft.	10in.
Well, after end	11ft.	0in.
Rowlocks	9ft.	1in.
Bulkhead.....................................	12ft.	0in.
Diameter of mast tube...................		3in.

In shape the new cruiser resembles the old sneakboxes, but is deeper than most of the latter. The board is of steel plate ¼in. thick, pivoted at the fore end in place of the dagger boards once common in these craft. The construction is quite peculiar. Both keel and planking are of white cedar ⅜in. thick. The keel is flat, in one piece, its half breadth being shown by the dotted lines in the half-breadth and body plans. There is no stem piece, but the keel is bent up, forming the stem. The garboards, also shown by the dotted lines, end along the gunwale, instead of in a rabbet in the stern, as in most boats.

In building, after the keel is fastened to the stocks, with the proper curve, the stern and moulds are put in place. Then two pieces, A A, are sawn out of 1in. board, the shape being taken from the deck line on the floor. These pieces are screwed in place, at the height of the lower side of the deck, and remain permanently in the boat. The ribs are now bent and fitted in place, nailed to the keel, and the upper ends of the forward ones are nailed to the pieces A A. As the planks are put on, they also are fastened at the fore end to A A. The correct breadths of each plank may be taken from the body plan on every frame. The frames are of sawn cedar 1¼x1¼in. and spaced 10in. The trunk is of pine, deck and ceiling ¼in. cedar. The rowlocks are of brass, fitted to fold down. A very peculiar feature of the new boat is a high washboard all around the gunwale, to keep off water and to serve as a receptacle for odd articles on deck. Forward the two sides are bolted to a chock of wood or an iron-casting of the shape shown, the top making a fair leader for the cable. Along the sides the washboards are held down by an iron catch, *a b*. The piece, *b*, made of band iron, ⅜x⅛in., is screwed to the deck and a notch in the lower side of the washboard hooks under it. The piece *a* is pivoted to *b*, serving, when closed, as a stop to keep the board in place. To remove the washboard, *a* is turned to one side, when the board may be slipped free of *b*. Aft the washboard is fitted with two battens, sliding into square staples in the stern.

A smaller boat of the same kind just built is 13ft. 5in. long. The centerboard, of galvanized iron, is placed 3ft. 8in. from stern, the trunk being 3ft. 1½in. long on bottom. The rig of the larger boat is a balance lug of the following shape:

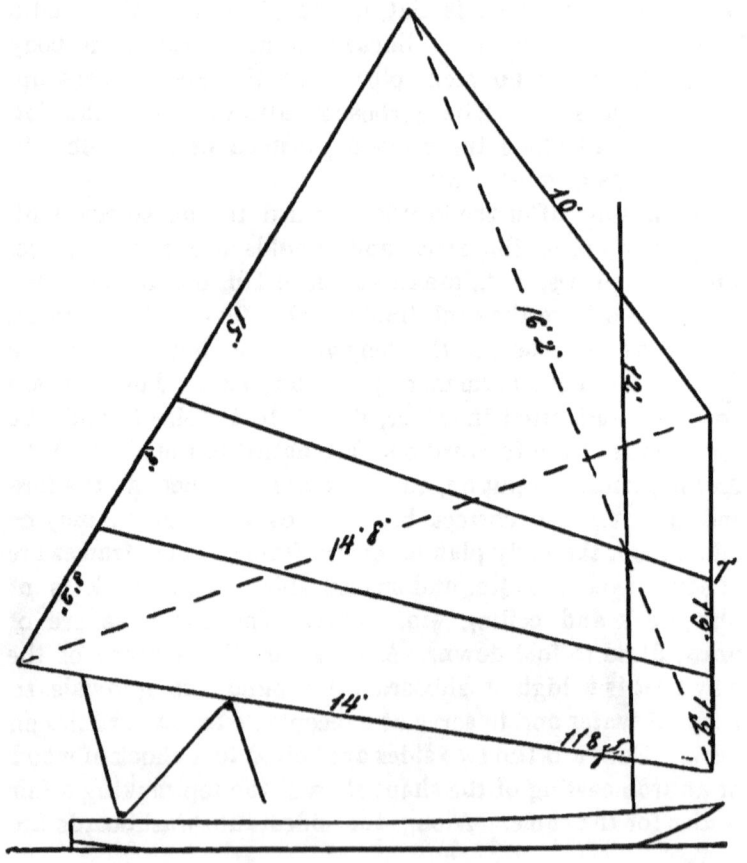

It has battens and is hung in the ordinary manner. A rudder may be used, but the boat is often steered with an oar.

The following description of the outfit of one of these boats is given by "Seneca":

Beginning at the stem, she is decked over 5ft. 10in. The centerboard trunk begins 3ft. 1in. from the bow and ends at the fore end of the cockpit. Between the stem and the centerboard trunk are an extra coil of rope and an extra coffee pot and tin pail. A shrimp net with handle and a jointed fishing rod also occupy part of this space, and extend part way alongside the starboard side of the trunk aft. To the starboard of the trunk, in easy reach of the cockpit, are two small oil stoves and a can of kerosene oil, also a brass rod which is used to shove down the centerboard. On the port side of the trunk are the clothes bag and the granite-ware cooking utensils, kettle, coffee pot, three cups and three plates. The cockpit tent is folded up on the floor close to the after end of the trunk, and next comes a tin water-tight box with the rubber bag of bedding atop of it, which is used as a seat when rowing.

The box is divided into compartments. No. 1 compartment contains awl, gimlet, screwdriver, nippers, oyster knife, cartridge loading tools, brass screws, screw eyes, brass and galvanized blocks, safety-pin hooks, nails, rings, spare cleats, tacks, etc. No. 2 compartment contains unloaded shells. No. 3, loaded shells. No. 4, fishing tackle of all kinds, small mirror, comb, thread and needles. No. 5 contains gun-cleaning tips, waste, rags and a bottle of gun oil. In the cover of the box a jointed cleaning rod is held by springs. In the rubber bedding bag are mosquito netting, two blankets, a quilt and a thick carriage robe, and perhaps an extra flannel shirt or two that can't be crowded into the clothes bag. Between this seat and the after end of the cockpit is a clear space in which to "work ship."

The after deck is 3ft. long, covering a 2ft. cuddy and a

foot of room below decks. In the latter space are stowed the two water jugs, a rubber inflatable mattress, a rubber coat and a macintosh-covered basket containing bread, pilot biscuit, cheese, etc. In the stern cuddy are canned soups, canned plum puddings, sardines, and other tinned edibles, potatoes or other vegetables in water-proof muslin bags; a candle lantern, riding light, and odds and ends of all descriptions. Underneath the side decks on either side of the cockpit are little shelves between every two deck braces. There are seven of these shelves on each side, which, numbered from the stern, are occupied as follows:

	Port.	Starboard.
1	Bag of shot...............	Bag of shot.
	Can of powder............	Revolver.
2	Soap, sponge.............	Pipe, tobacco.
	Whisk broom.............	Box of matches.
	Scrub brush	
3	Monkey wrench...........	Hatch padlock.
	Can opener, big...........	Case-knife, fork.
	Spoon, pliers..............	Three teaspoons.
4	Coffee can................	Sugar can.
	Salt can...........	Condensed milk.
	Pepper box...............	Bottle chow-chow.
5	Generally vacant, the bedding bag preventing easy access.	
6	Spare rowlocks....	Hatchet.
	Spare blocks.	
	Candles.	
7	Towels	Grub in general.
	Slippers.	

The gun lies on the floor under starboard side deck and the skipper's artificial aids to walking under port side deck. On deck, between stem and mast, 2ft. 9in., is coiled the anchor cable, with Chester folding 12lb. anchor. On side decks, where the 6in. high washboard prevents their rolling off, are the oars, boathook, mast and sail when not in use. A stern cable is coiled on after deck.

In sailing a long-handle tiller is used, so that steering can be done from the cockpit, but under certain conditions the skipper steers from the after deck, with the tiller put on the rudder head "stern foremost," the handle sticking out astern like a boomkin. The cruising sail generally used is a spritsail, which can be stowed below, the hatches put on and locked, and the cruiser left at any port with everything in her, while the skipper takes the train home to spend Sunday with his family.

With such arrangements as the above the skipper lives aboard his boat, sometimes not touching shore for three or four days. Sitting on her oilcloth covered floor to cook a meal, he can reach everything necessary without moving his position; sitting there at night with the tent up he has 4ft. of headroom in a waterproof cabin, which can be made warm and cosy in December by keeping one of the oil stoves alight; and anchored at night in a cove he sleeps like a top on a soft "air mattress," rocked gently by the waves.

PLATES XL. AND XLI.—"FOREST AND STREAM" CRUISER.

The sneakbox is essentially a hunting boat, and the Barnegat cruiser shown in Plate XXXIX. partakes largely of the same characteristics. As the attention of boating men has been more generally drawn to the cruising qualities of the sneakbox, many comments, criticisms and suggestions have been made for the improvement of the model as a cruiser, leaving out all considerations of duck shooting and looking only to the end of a safe, speedy and convenient boat, adapted both for general cruising on open water and a safe boat for summer sailing. To meet these wants the accompanying design was made and a boat built which has proved very satisfactory on trial. The new craft is based on the sneakbox, the bottom of which is kept almost intact; but an inspection of the former boat showed several features capable of alteration, if cruising only was considered. In the first place, the low sides, excellent if the boat is to be used as a blind, have been built up; the excessive crown of deck has been reduced, and the washboards have been discarded. As the height of the deck in the new boat is less than that of the washboard on the old, the windage is reduced, while the room inside and the stability are both increased by the additional bulk of the new boat. At the same time the new boat will stow for transport in the same breadth and height as the old, the total depth being the same. The increased freeboard and higher bow improve the boat greatly in rough water.

On the other hand, the high washboards made a convenient receptacle for the oars, etc., but the extra inside room in the latter offers a full compensation. The folding rowlocks are given up entirely, thus removing a troublesome appendage, and cleats are fitted to the coaming, in which ordinary socket rowlocks are set. If it is desired to use a longer oar the cleats may be screwed to the deck near the gunwale. As there is no special virtue in the awkward-looking square stern of the sneakbox, the deck and planking have been extended 2ft. aft, the latter merely continuing in a fair upward curve until they meet at the gunwale as in the bow. This gives a handsome finish to the boat, in the shape of an elliptical stern, with an easier run, more buoyancy and increased deck room. The rudder is of the balanced variety, a suggestion of the owner of the Bojum, the stock being of $\frac{3}{4}$in. iron, to the lower end of which two flat pieces 1×$\frac{1}{4}$in. are welded, making a shape like a tuning fork. In this fork a piece of 1in. oak is set, forming the rudder, the head of the stock is squared for a tiller, and at the level of the deck a hole is drilled for an iron pin, supporting the whole. To form the rudder trunk a piece of pine 3in. square is fitted from the inside of the planking to the deck, being set in white lead and well screwed through plank and deck. Through the center of this piece a vertical hole 1in. in diameter is bored for the rudder stock.

Owing to the extended deck aft, both the cockpit and centerboard are further aft than in the ordinary sneakbox. The coaming of the cockpit is 2$\frac{1}{4}$in. high. The floor boards are raised from 3 to 3$\frac{1}{2}$in. above the bottom, so that the bilge water will not slop over them, the extra depth allowing this change. There are no fixed thwarts, the oarsman sitting on a box which holds the stores, etc., on a cruise, while in sailing the crew sit on deck or on the floor. When used for pleasure sailing five or six may be accommodated, and in cruising a bed for three can be made up on the wide floor of the 16ft. boat. A tent can easily be rigged over the boat at night, supported by the boom. For one or two persons such a boat 13ft. over all will be quite large enough for cruising, and may be built of light weight. The dimensions and scales are for two sizes, 16 and 13ft. over all.

TABLE OF OFFSETS—THIRTEEN-FOOT CRUISER.

Stations	Heights		Half-Breadths							
	Keel	Deck	Deck	No. 1	No. 2	No. 3	No. 4	L W L	No. 6	No. 7
	Ft.In	Ft.In	Ft.In	Ft.In	Ft.In	Ft.In	Ft.In	Ft.In	Ft.In	Ft.In
0	1 7^4	1 7^4	0							
1	1	1 6	7^3							
2	5	1 4^6	1 0^3	9^6	8^4	6^4	3^7			
3	1^4	1 8^4	1 4^1	1 3	1 2^2	1 1^1	11^5	9^6	6^7	0^5
4	0^2	1 2^6	1 6^6	1 6^2	1 5^5	1 4^7	1 3^7	1 2^4	1 0^6	9^5
5		1 2^1	1 8^5	1 8^1	1 7^6	1 7	1 6^1	1 4^6	1 3^1	1 0^2
6		1 1^7	1 9^5	1 9^2	1 8^6	1 8	1 7^1	1 5^6	1 3^7	1 0^7
⊠		1 1^6	1 10	1 9^5	1 9^1	1 8^3	1 7^4	1 6	1 4	1 1
8		1 1^6	1 0^6	1 9^3	1 9	1 8^2	1 7^2	1 5^6	1 3^7	1 0^6
9		1 1^7	1 9	1 8^5	1 8^2	1 7^4	1 6^4	1 5^1	1 3	11^2
10	1^2	1 2^1	1 7^4	1 7	1 6^4	1 5^5	1 4^4	1 2^5	11^3	
11	3^7	1 2^4	1 5	1 4	1 3^2	1 1^5	10^7	6^4		
12	8^2	1 2^7	1 1^3	0^6	6^4					
13	1 3^4	1 3^4								

TABLE OF OFFSETS—SIXTEEN FOOT CRUISER.

Sta.	Keel.	Deck	Deck	No. 1.	No. 2.	No. 3.	No. 4.	L W L	No. 6.	No. 7.
0....	2	2	0	......	......	......	......	......	......	......
1....	1 5^2	1 9^4		7^6	......	......	......	......	......	......
2....	1 0^3	1 9	1 1	8^5		6^3	3	......	......	......
3....	5^2	1 7^6	1 5^3	1 3^1	1 2	1 0^4	10	6^3	......	......
4....	1^2	1 6^6	1 8^6	1 7^4	1 6^5	1 5^5	1 4	1 1^7	11	5^3
5....	0^1	1 6	1 11^3	1 10^4	1 9^7	1 8^7	1 7^6	1 6	1 3^7	1 0^2
6....	......	1 5^3	2 1^1	2 0^4	2	1 11	1 10	1 8^3	1 6^5	1 2^6
7....	......	1 5^1	2 2^2	2 1^6	2 1^2	2 0^2	1 11^2	1 9^5	1 7^3	1 3^6
8....	......	1 5	2 2^6	2 2^3	2 1^6	2 0^6	1 11^6	1 10	1 7^5	1 3^7
9....	......	1 5	2 3	2 2^5	2 2	2 1	2	1 10^1	1 7^5	1 3^7
10....	......	1 5	2 2^7	2 2^2	2 1^6	2 0^6	1 11^5	1 9^6	1 7^4	1 3^4
11....	......	1 5^1	2 2	2 1^3	2 0^7	2	1 10^6	1 9	1 6^4	1 2^1
12....	1	1 5^2	2 0^4	2	1 11^2	1 10^4	1 9	1 7^1	1 3^6	9^4
13....	3^1	1 5^4	1 10^3	1 9^4	1 8^6	1 7^3	1 5^3	1 1^7	7^5	......
14....	6^4	1 5^7	1 7^4	1 5^5	1 4	1 1	8	......	......	......
15....	11^4	1 6^3	1 3		9^1	2^4	......	......	......	......
16....	......	1 7	0	......	......	......	......	......	......	......

DIMENSIONS OF "FOREST AND STREAM" CRUISER.

Length over all..................	13ft.	16ft.
waterline...............	9ft. 4^2in.	11ft. 7in.
Beam, extreme..................	3ft. 8in.	4ft. 6in.
Depth at gunwale.............	1ft. 1^6in.	1ft. 5in.
Sheer, bow......................	5^6in.	7in.
stern	1^6in.	2in.
Crown to deck..................	2in.	3in.
Fore side of stem to—		
Mast.......................	2ft. 5^4in.	3ft.
Trunk, fore end............	2ft. 8in.	3ft. 3in.
after end..........	6ft. 4in.	7ft. 9in.
Well, fore end.............	5ft. 8in.	6ft. 11in.
after end..........	10ft. 11in.	13ft. 6in.
Rowlocks..................	8ft. 7in.	10ft. 6in.
Rudder......................	11ft. 11in.	14ft. 8in.
Width of well..................	2ft. 6in.	3ft.

The stations are 1ft. apart by both scales. In the 13ft. boat the waterlines are 1⅜in. apart, and in the 16ft., 2in. The scantling for the 13ft. boat would be, planking ¼in., deck ⅜in., timbers ⅜×½in., spaced 9in. The larger boat would have ⅜in. planking, ½in. deck, and timbers 1×½in., spaced 10in. The stern is framed as described for the stem on page 182, two quarter pieces being cut to the outline of deck and fastened to transom and upper end of keel. A sternpost and scag are fitted after the boat is taken from the stocks, and two bilge keels are screwed outside. The centerboard is of yellow pine, edge-bolted with ¼in. iron and weighted with lead. The deck is covered with 6oz. duck, laid in fresh paint. A half round bead makes a finish around the gunwale and covers the edge of the canvas. If a handsome little sailing boat is desired, the hull above water will be painted black or white, with a gold stripe as shown, the bottom being coated with copper bronze. The boat shown was built by J. MacWhirter, of West New Brighton, Staten Island. The cost will vary according to size and finish, from $125 for a 13ft. boat with sail and galvanized fittings to $160 for a 16ft. boat finished with brass fittings.

Only three forms of sail are in common use in American waters, the boom and gaff, the leg of mutton or sharpie, and the sprit, and of these the former is by far the most common. In spite of its serious disadvantages, and the fact that there are many better rigs, it has held its own for many years, and is still as popular. Within a half dozen years the canoe men have given to the world a number of new rigs, either of new design or adopted from abroad, and in this point of good and efficient sails, these new sailors are far ahead of the older boat-sailing experts with far more experience. Chief among the new-fangled ideas of the canoeist is the balance lug, an English adaptation of a Chinese sail, now extensively used in this country and applicable to all small boats. This sail has been chosen for the "Forest and Stream" cruiser, and has worked very successfully.

Of course the first requisite in going to windward is a taut luff, as with it shaking nothing can be done. With a badly cut and made lug sail this cannot well be had; but a boom and gaff sail has this doubtful advantage, that by means of two halliards it may be strained and stretched into some kind of shape, though never what it should be. With a properly cut sail this advantage in favor of the boom and gaff disappears.

On a small boat one sail, if rigged so as to be easily handled, is not only faster, but much more easily managed than two, one being a jib. It is of course much better to windward or free, while there are fewer lines. The requisites for such a sail are different in a large and in a small boat, as in the former there is much more room to stand and work halliards and lines; there are usually more to help, and the mast is always kept standing. In a small boat the sail must hoist and lower easily, surely and quickly; it must be readily removed from the mast for stowage or in rowing, and it must be so placed as to balance properly in connection with the keel or centerboard. In all of these particulars the sail shown is better for sneakboxes, yachts' yawls, rowing and sailing boats, and other small craft, than the boom and gaff. The former has no mast rings to jam in hoisting and lowering, as they are always liable to do; it can be quickly removed from the mast; the latter is stepped much further from the bow, keeping the weight aft and being easily reached and unstepped, while before the wind the sail is not all on one side of the mast and boat, but a large portion is so placed as to help balance the outer end.

The sail shown is for the 16ft. cruiser, and is made of yacht drill, 6oz., double bighted, bights running parallel with the upper portion of the leach. The gear is rigged as follows: The boom, 2in. greatest diameter, is 14ft. long, to allow for stretch, and is laced to the foot of the sail, the latter having about 3in. roach or rounding. A single brass block (i) is lashed to the outer end for the sheet (f). Just abreast of the mast is lashed a snaphook. As the greatest strain on the boom is at this point, it is stiffened by a fish batten (*l l*) of oak, ½in. square at middle and tapering at ends, the length being 2ft. This batten is lashed to the boom by four lashings of fine twine, and adds materially to the strength, while lighter and less clumsy than an enlargement of the fore end of boom would be.

The head of the sail is cut with a round of 9in., 1in. per foot, for the following reason: A straight stick, like a yard, is very elastic, even if of considerable size, and will bend greatly at the ends. If, however, it be curved in the first place, it then requires some force to bend it further. The principle is well shown in the common bow, which is easily strung, but then requires a heavy pull to bend it. Another important advantage follows this form; the yard or bow is first curved in a vertical plane and held there by the sail. Now, with this tension on it, it resists powerfully any lateral strain that would throw the peak to leeward. This is aided by the peculiar cut of the sail. The yard is brought far down the luff and a large part of its length is forward of the mast. When the luff is properly set up a very strong leverage is put on the yard, holding the head well to windward. The sail is approximately square in shape, as this form gives the maximum area with a minimum average of spars, mast, boom and yard. The clew is cut off, as will be seen, as a shortening of the yard by a foot or so lessens the area but little. Two battens are placed in the sail as shown, with reef points, and a hand reef may be added, such as has been described previously for canoes. It will be simpler to run the hand reef to a cleat on the boom instead of on deck, as a man can stand up readily in a large boat, and can reach the boom near the fore part, while in a canoe he must keep his seat, consequently the lines must lead to his hand, at the cost of simplicity. The battens are $1\frac{1}{4} \times \frac{3}{4}$ at middle and ¾in. square at ends, and are run in pockets in the sail.

The yard is 1½ diameter at largest part, the middle third of its length, and is rigged as follows: A rope strap (*o*) is worked on it, a fish batten (*m m*) being used as on the boom. The eye of the strap is large enough to admit a snaphook on halliard, or better yet, a snatchblock may be employed. The halliard (*a a*) of 9-thread manilla rope, leads through a sheave at masthead, thence through a deck pulley near mast, and is belayed on one of two cleats on deck at the fore end of well. In its upper end a snap or gafftopsail hook is spliced, and on the mast is a 4in. galvanized iron or brass ring (*n*) bent into oval form, 3½×4¾in. It must be large enough to slide readily without danger of jamming. The halliard is passed through the strap on yard and then hooked to the ring. When hauled taut the yard is always held in to the mast, whether full sail or reefed is carried.

A lug sail can hardly be set taut by a halliard, but a tack tackle must be employed, and a very powerful one is rigged as follows: On the mast is another ring, to which is lashed a brass block (*d*). On deck is a deck pulley at port side of mast. The tack line (*b*) is made fast to the deck abaft the pulley, the end is rove up and forward through the block (*d*), thence down and aft through deck pulley to cleat at fore end of well. In setting sail the mast is stepped, hook on boom is snapped into the eye of block (*d*), the two parrels on the bottom are tied, the halliard is passed from aft forward through the strap on yard and hooked to upper ring (*n*); then the sail is hoisted as high as possible, after which, when the halliard is belayed, the tack is hauled down until the sail is perfectly flat. In reefing or lowering it is best to start the tack first, then when the halliard is set up the tack is hardened down again.

A toppinglift (*e*) is thus fitted: The line is double, running from masthead down each side of sail and splicing into one just below boom, leaving enough slack to lower the latter. On the boom is a fairleader (*k*) lashed fast, and through this the toppinglift is rove, thence to a cleat on boom. It may thus be easily reached for a pull at any time, even with the boom hard off. In hoisting or reefing the toppinglift should take the weight of the boom always.

In removing the sail the end is cast off and the lift remains on the mast. In setting sail the latter is first dropped into the bight of the toppinglift, the fore end of spars on deck at port side of mast. The lift is made fast to cleat, raising the sail, the tack and halliard are snapped on, and all is ready for hoisting. A jackstay, from masthead, and made fast to mast about 1ft. above deck, will be found very useful in holding up the fore ends of spars and sail.

On each batten a parrel is made fast, to hold the sail to the mast on the starboard tack. These are small lines about 2½ft. long, the fore ends fast to the battens, while the after ends are tied or hooked into rings lashed to the battens.

The sheet in a small boat is always a trouble, wherever it may be made fast it is always in the way. The plan adopted in the present case is perhaps as good as any. The sheet runs through a block (i) on boom and an eye splice is worked at each end. On deck are three cleats, one (g) just abaft the rudderhead, and the others (h) on each side of well. When on the wind the after eye of sheet is hooked over the after cleat (g) while the sheet is held in the hand or belayed to one of the forward cleats; or if desirable the eye may hook over one of the latter. When the boom is off the sheet is cast off from the cleat and its whole length is used, the eye at the end preventing it from unreeving from the block. By this method a very short sheet is required, while a good purchase may be had when on the wind, and the sheet can always be arranged to be out of the way of the sailor.

Fitted as described the sail will be found a very effective one, and once accustomed to it there is no difficulty in handling it quickly, while it is much less troublesome and cumbrous than a mainsail and jib. If for any reason the latter must be used, it can be fitted on a stay, the luff of the sail being cut down as much as possible, so as to allow room for the jib, but in almost all small boats the single sail will be found best.

PLATES XLII. AND XLIII.—SAILING AND CRUISING BOAT "DELTA."

The many inquiries concerning sneakboxes, tuckups, small sharpies and similar craft show that there is a general demand for small sailing craft of good design, both for ordinary pleasure sailing and for more or less extended cruising. This demand may be largely ascribed to the influence of canoeing, as it has arisen since the latter sport became generally popular in this country. There are many to-day who have given up the canoe after a fair trial, and many more who are ready to do so; but this is not on account of any defect in the boat. The general popularity of canoeing, together with the moderate expense, leads many novices into it, not because it is just the form of sailing which they prefer, but because they know of no other which would suit them better. After a time some find the canoe too small to carry a party of friends, others wish a boat exclusively for sailing in open waters, and others, again, wish to carry an amount of stores, guns and tackle, for which the canoe never was intended. This proves nothing against the canoe, a boat adapted to wider range of use than any other pleasure craft; but when a man gets to this stage and begins to long for a sneakbox or a sharpie, he is better out of a canoe than in it, though there is no reason why the bond between him and the man who still swears by a 15×30 canoe should be severed; they are both cruisers and sailors at bottom, though their craft may vary.

The boat shown in the accompanying plans, the Delta, was planned by Dr. H. G. Piffard, former owner of the sneakbox Bojum(Pl. XXXVIII.), and is an attempt to combine the best qualities of several boats. The bottom of the sneakbox is preserved, but with the bows of the ordinary rowboat, as well as a higher side, while the overhang and rudder of the sharpie are added. The boat was intended for pleasure sailing about Greenwich, Conn., to carry half a dozen comfortably, and yet to be easily handled by one. While a fair amount of speed was looked for, the boat was not intended for racing, and if wanted for such a purpose, to carry all the sail the model is capable of with a crew on the gunwale, a heavier construction would be advisable. For all ordinary work the boat has proved amply strong, and the construction here given can be followed in all details.

The question has often been asked, "Why not put a boat bow on a sneakbox?" and for all save hunting purposes there seems to be no reason why it should not be done, in fact this boat is a practical answer to the question. The Delta was built in the spring of 1886, and thus far has given perfect satisfaction. In order to meet the wants of the single-hand cruisers the drawing is given with two scales, by which a boat of 13ft. extreme length may be built, as well as the orignal length of 18ft. The former should make a remarkably good little craft, larger, faster, abler and far handsomer than the sneakbox, and little more costly. The bow is not so well adapted for beaching, and the boat is too large and high to serve as a blind or shooting battery, as a sneakbox often does, but as far as sailing and general cruising are concerned the odds are all in favor of the Delta as compared with any form of "box." The dimensions of the two sizes are:

	18FT. BOAT.	13FT. BOAT.
Length over all	18ft.	13ft.
l.w.l.	16ft.	11ft. 6^{6}in.
Beam, extreme	5ft. 4 in.	3ft. 10^{1}in.
Draft, about	8 in.	6 in.
Depth at gunwale, amidship	1ft. 5^{4}in.	1ft. 0^{5}in.
Sheer, bow	7 in.	5^{1}in.
stern	3 in.	2^{1}in.
Crown of deck	2 in.	2 in.
Fore side of stem to—		
Trunk, fore end of slot	4ft. 9 in.	3ft. 5^{2}in.
after end of slot	9ft. 6 in.	6ft. 10^{4}in.
Well, fore end of slot	8ft. 9 in.	6ft. 4 in.
after end of slot	15ft. 9 in.	11ft. 4^{5}in.
Rudderstock, center	16ft. 8 in.	12ft. 0^{4}in.
Rowlocks, center	11ft. 7 in.	8ft. 4^{4}in.
Width of well	4ft.	2ft. 10^{4}in.
Distance of stations apart	2ft.	1ft. 5^{2}in.
waterlines apart	3 in.	$2^{3}{}_{16}$in.

TABLE OF OFFSETS—EIGHTEEN-FOOT BOAT.

Stations	HEIGHTS.		HALF-BREADTHS.					
	Deck.	Keel.	Deck.	No. 1.	No. 2.	No. 3.	No. 4.	Keel.
0..	2 0^4	....	0^1	0^1	....	....	....	....
1..	1 10	....	1 1^5	11^2	9^7	8^2	5^7	1^2
2..	1 7^5	....	1 11^2	1 8^6	1 7^2	1 4^5	1 0^5	1^6
3..	1 6^2	....	2 4^7	2 3^3	2 1^7	1 11^4	1 6^5	2^3
4..	1 5^4	....	2 7^4	2 6^4	2 5^1	2 3	1 11	2^5
5..	1 5^4	....	2 8	2 6^7	2 5^7	2 3^6	2 00	2^6
6..	1 5^6	0^3	2 7	2 5^6	2 4^5	2 1^6	1 9^3	2^7
7..	1 6^4	2^4	2 4^5	2 2	1 11^7	1 7^2	7^3	2^7
8..	1 7^3	9	1 11	1 3^6	....	....	....	2^7
9..	1 8^4	1 8^4	....	....	....	....	....	2^6

TABLE OF OFFSETS—THIRTEEN-FOOT BOAT.

Stations	HEIGHTS.		HALF-BREADTHS.					
	Deck.	Keel.	Deck.	No 1.	No. 2.	No. 3.	No. 4.	Keel.
0..	1 5^6	....	0^1	0^1	....	....	....	....
1..	1 3^6	....	9^6	8^2	7^2	6	4^3	1
2..	1 2^1	....	1 4^6	1 3	1 1^6	1	9^2	1^2
3..	1 1^1	....	1 8^7	1 7^6	1 6^5	1 5	1 1^5	1^6
4..	1 0^5	....	1 10^5	1 10	1 9	1 7^4	1 4^5	1^7
5..	1 0^5	....	1 11^2	1 10^3	1 9^4	1 8	1 5^3	2
6..	1 0^6	0^1	1 10^2	1 9^4	1 8^4	1 6^5	1 3^4	2
7..	1 1^2	1^7	1 8^5	1 7	1 5^2	1 1^7	5^3	2
8..	1 1^7	6^4	1 4^5	11^3	....	....	....	1^7
9..	1 2^6	1 2^6	....	....	....	....	....	1^6

The scantling for the larger boat will be: Keel of oak, ⅞in. thick and 5in. wide; stem of hackmatack or oak knee, sided 2in.; sternpost, oak, 1×2in.; scag, yellow pine, 1in. thick; planking, cedar, ¼in., the garboard ⅜in.; deck, cedar or white pine, ¼in. scant; coaming, oak, ⅜in.; deck beams, 1¼×1¼; ribs, 1×⅜in., spaced 9in. A clamp, 2×⅝in. at middle, tapering to 1½×⅝in. at ends, will be run inside from the bow to the bulkhead, being riveted through the ribs and upper streak. It should be set so far below the gunwale as to allow the deck beams to rest on it. A solid chock should be fitted in place of a breast-hook at the bow, below the deck and on top of these clamps. The after bulkhead will be of 1in. pine or spruce. The well for the centerboard will have headledges, *b b*, of oak, 1×2in., with bed pieces, also of oak, 3½in. deep and 1¼in. thick, the sides above being of 1in. clear white pine.

The keel is ⅜in. extreme thickness, but it may be tapered, beginning at station 5 and reducing it to ¼in. at the after end, so that it will bend more easily. It will be thicker than the garboards, but when the planking is completed the bottom will be planed down on the edges to meet the latter.

The smaller boat will have keel ⅜in. thick; stem, sided 1¼in.; ribs, ¾×⅜in., spaced 8in.; planking, ¼in.; deck, ⅜in.; coaming, ⅜in. scant; deck beams, 1in. wide and 1¼in. deep; headledges, 1½×⅜in.; bedpieces, 3×1¼in. In construction, the keel is first laid on the stocks and the stem is got out and fitted. A mould is made for every station, that for station 8 being carefully beveled and fitted, as it is to remain permanently as a bulkhead. A mould is now made of common stuff to fit the upward curve of the keel from station 5 to the stern, and is set up on the stocks, the keel being shored down into place. The moulds, eight in number, including the bulkhead, will then be set in place. Only half as many moulds would be used by a regular builder, but the amateur will find the work easier if he has plenty of moulds. The shape of the stern is given by two pieces, *ff*, termed quarter timbers. These need be only of 1in. pine or spruce. They must be marked out from the lines on the floor and carefully beveled. They are screwed to the keel and also to the bulkhead, being let into the latter.

A number of ribbands of oak or yellow pine, with clear straight grain, are now run around the moulds, about six on each side. The ribs are now planed up, steamed, and bent into place, being held by a nail partly driven through each ribband. Each rib is long enough to lap the full width of the garboard, the two that go to make a frame lying side by side where they cross. When the ribs are all in and fastened permanently to the keel and temporarily to the ribbands, the lower ribband on each side is taken off, the garboard got out and set. The next ribband is then removed and another plank is set and so until the boat is planked. The piece a is of oak, $1\frac{1}{4}$in. thick, set down on the keel to stiffen it and form a mast step. It should be put in place before the ribs go in, the latter being jogged in and well fastened to it. The clamps should be put in before the frame is taken from the stocks, the deck beams being also fitted.

The slot for the trunk should never be cut until the last thing, but when the planking is completed, two mortises are cut for the headledges, the two bed pieces are got out and fastened to them with through rivets, the boat is taken from the stocks and the trunk put in, screws being driven through the keel into the bed pieces. The sides of white pine are then put on, after which the deck beams, previously fitted, will be fastened. They should run across the trunk, being jogged down so that the deck will close the top of the trunk tightly. The mortise for the sternpost is next cut, the post put in and fastened to the bulkhead, then the scag is cut and fitted, being fastened through with screws from inside of keel. To make a tight casing for the rudderstock, a piece of pine $3\frac{1}{4}$in. square, e, is fitted to the keel, reaching to under side of deck, being set in whitelead and well screwed to make a watertight joint. A hole is then bored for the iron rudderstock. Ledges about 3in. deep run across the boat to carry the floor boards, and also to stiffen the bottom, for which purpose they should be well riveted through the planking.

The decks are supported along the well by knees $c\ c\ c$, three or more on each side, forming lockers. One or more of these may be fitted with doors as shown. After the boat is taken from the stocks the slot for the centerboard is cut in the keel. The rudderstock is shown in detail, the shank of round iron, $\frac{3}{4}$in. with two flat pieces each $1 \times \frac{1}{4}$in. welded to it, the head being squared for a tiller. The blade of the rudder is of oak, 1in. thick where it is let into the stock, but tapering to a fine edge forward and aft. On deck there should be a brass plate, while a pin through the stock prevents it from falling. A strip of iron $\frac{3}{4} \times \frac{1}{4}$ may be run from the centerboard slot aft to the rudder, with a pin up into the center of the latter, as shown. This will serve the double purpose of stiffening and protecting the scag and rudder, and also of preventing the fouling of the latter by weeds and lines. The deck should be covered with light drill, 6oz. laid in white paint. All fastenings should be of copper except where brass screws are used. The sizes given for planking and decks are for planed stuff, and in all cases are thick enough to allow of planing off after caulking, which will always be necessary. The board shown is large enough for all sailing, and in the smaller boat, if room is an object, it might be shortened by 6in. on the after end. The size of cockpit may be varied according to the boat; for a large party it may run further forward on each side of the trunk, but for rough water and cruising the size shown will answer very well. The mortise for the mast step may be cut in the piece a. The forward deck should be strengthened by a piece $\frac{3}{8}$in. thick and 6 to 7in. wide, running from stem to trunk under the deck beams and riveted through beams and deck plank. Where the mast goes through the space between, the two should be filled in solid.

The sail shown in Plate XLI. is well adapted for the smaller boat and will be none too large, but in Plate XLIII. two other suitable rigs are shown, the scales given being adaptable to either size of boat. Delta was rigged with a single large lug of about 150ft., which she carried easily without ballast in ordinary weather. The lug was cut with very little round to the head and the yard was straight. There were no battens in the sail, but two rows of reef points. The sheet was fast to an eye bolt on the quarter and led through a block on the boom, and then under a thumb cleat on the side of coaming, or through a snatch block on the floor of the boat. Under this rig the boat handles very satisfactorily for singlehand sailing, but some ballast would be needed if no passengers were carried. The dotted lines show the size of a cat rig, the mast being stepped further forward. The lug is the better of the two, but is more difficult to rig, and many will prefer the cat simply because they are used to it and unfamiliar with the other. The main and mizen rig in the second drawing is smaller, and better adapted for cruising and sailing alone. The details of the rigging are the same as in the preceding sail plan. The sails here shown are of the form usually carried on canoes and small boats, but a flatter and more effective sail can be had by making the yard longer, carrying it down to the batten, rounding the head much more than is shown, and throwing a little more of the yard forward of the mast, as in the Cruiser's sail. Such sails require to be carefully rigged, and more or less trial is always needed to find the best position for halliard and tack on the spars, but when once complete they are good enough to fully repay the trouble. The dimensions of the single lug are as follows, the mast in the plan being shown forward of its proper position, which is given in the table:

	18ft. boat.	18ft. boat.
Mast, from stem	4ft.	2ft. 9in.
above deck	15ft.	10ft. 10in.
diameter at deck	3¼in.	2¾in.

AMATEUR CANOE BUILDING.

	18FT. BOAT.	13FT. BOAT.
Boom	14ft. 9in.	10ft. 6in.
diameter	2in.	1¾in.
Yard	10ft. 6in.	7ft. 7in.
diameter	1¼in.	1⅛in.
Foot of sail	14ft.	10ft. 1in.
Luff	9ft. 6in.	6ft. 10½in.
Head	9ft. 9in.	7ft.
Leech	19ft.	13ft. 8in.
Tack to peak	18ft. 6in.	13ft. 4in.
Clew to throat	10ft. 6in.	11ft. 11in.
Area	155sq. ft.	80sq. ft.

The dimensions of the main and mizen rigs are:

	18FT. BOAT.		13FT. BOAT.	
	Main.	Mizen.	Main.	Mizen.
	Ft. In.	Ft. In.	Ft. In.	Ft. In.
Main, from stem	2 00	17 00	1 06	12 03
above deck	12 06	7 00	9 00	5 00
diameter at deck	0 03	0 02	0 02¼	0 01½
Boomkin, outboard		1 09		1 03
Boom	12 02	6 06	8 10	4 09
diameter	0 03¾	0 01½	0 01¼	0 01
Yard	9 06	5 03	7 00	3 10
diameter	0 01¼	0 01	0 01½	0 00⅞
Foot	11 06	6 00	8 04	4 04
Luff	6 06	3 06	4 09	2 07
Head	9 00	4 00	6 06	3 06
Leech	14 06	7 07	10 06	5 06
Tack to peak	15 00	8 00	10 10	5 09
Clew to throat	12 06	6 06	9 00	4 09
Area	90sq.ft.	26sq.ft.	40sq.ft	14sq.ft.

The drawings show the size of sails when stretched on the spars after a little use. They must be cut a little smaller than this in making, and after a season they will have stretched to the full size of the spars shown.

PLATE XLIV.—THE DELAWARE RIVER DUCKER.

Along the Delaware these boats are much used for rowing and sailing generally, gunning and fishing, but especially for redbird shooting in the marshes below Philadelphia. The flat floor allows them to be poled far up on the marshes where there is more mud than water, and they are often propelled by a long pole with three prongs on the lower end, for poling on muddy bottoms. There is no fixed thwart, but a movable box is used, so that the gunner may sit in the fore end and his assistant may use the pole in the after end; the latter may sit forward and row while the gunner occupies the after seat; or the boat may be backed down by the oarsman in the after seat, the gunner sitting on the box in the bow. Both ends of the boat are exactly alike, the only difference being in the seat, rowlocks and coaming.

The dimensions are: Length, 15ft.; beam, 3ft. 10in.; depth, 13in.; sheer, 8in. The stem and stern are sided $1\tfrac{1}{2}$in., keel sided 6in. amidships and moulded 1in.; planking, $\tfrac{3}{8}$in.; timbers, $\tfrac{3}{8}$in.$\times\tfrac{5}{16}$in.; deck, $\tfrac{3}{8}$in.; flooring, $\tfrac{1}{4}$in.

TABLE OF OFFSETS—DELAWARE RIVER DUCKER.

Stations.	Deck Height.	Half-Breadths.				
		Deck.	12in.	9in.	6in.	3.n.
1.........	21^2	0^2				
2.........	18^4	10	7^6	6^1	4^2	1^6
3.........	15^2	18^2	16^6	15	12	7^6
4.........	13^5	22	21^6	20	17^4	13
5.........	13	23	23	21^6	19^4	15^2
6.........	13^5	22	21^6	20	17^4	13
7.........	15^2	18^2	16^6	15	12	7^6
8.........	18^4	10	7^6	6^1	4^2	1^6
9.........	21^2	0^2				

The stations are spaced 2ft. apart, measuring each way from midship section, and the waterlines are 3in. apart.

Along the bottom of keel are two wooden runners, A A, $\tfrac{3}{4}\times\tfrac{3}{8}$in. and shod with half-round iron. It will be noticed that the stem and stern each project the same distance below the planking, and the runners shown by dotted lines in the breadth plan project forward of the stem and aft of the stern, as at A A, B B. The floor boards are screwed to two battens, which are on top of them, so as to allow the boards to lie close to the bottom of the boat. They form one piece only, that may be easily lifted out. The brass rowlocks are accurately turned and fitted, with long shanks, so as to be nearly noiseless. The side decks are supported by three iron knees on each side.

The ducker carries one boom and gaff sail; the usual area for a boat of this size being 112sq. ft., the racing rig running up to 150sq. ft. The smaller sail would have 15ft. on foot, 7ft. 6in. luff, 16ft. 6in. leech, and 7ft. 6in. head, the mast being stepped 2ft. from stem.

PLATES XLV. AND XLVI.—DELAWARE RIVER TUCKUP.

The Delaware River in the vicinity of Philadelphia is the home of three special classes of sailing boats, the hiker, the tuckup and the ducker, all three being peculiar to the locality and used so far as we know on no other waters than the middle Delaware and lower Schuylkill. All are cat rigged, but differ in size, the hiker being the largest, a small half open catboat, with about 4ft. 8in. beam for 15ft. length, same proportion for larger sizes; the tuckup being a few inches narrower and not quite so deep, both with square sterns, while the ducker is a double-ended shooting skiff, also fitted for sailing. The plans show a very good example of the present fourth class tuckup, the Priscilla, built in 1887 for Mr. Edward Stanley, of Bridgeport, Pa., by James Wignall, of Philadelphia. The lines were carefully plotted from offsets taken by Mr. E. A. Leopold, of Norristown, Pa., the boat being enrolled in the Montgomery Sailing Club of that place and sailing in all the races. The peculiar name "tuckup" is derived from the fact that in building the flat keel is not carried out straight from the stem to sternpost, along the finely dotted line B, as would be the case in most catboats of any size, but it turns or "tucks" up, in builder's parlance, to the height of the waterline, as in the Delta, 'Forest and Stream" cruiser, and the sneakbox; a separate scag being added below the keel. The word came into general use from the construction and is now applied solely to such a boat as is here shown.

The two boats here described, Priscilla and Igidious, are owned on the Schuylkill about fifteen miles from Philadelphia, and sail in the races, but the home of the craft is in the Kensington district of Philadelphia, near the famous Cramp's shipyard. Here there are several long wharves, lined on each side with rows of two-story boat houses, twenty to thirty in a row. In these houses are stored hundred of duckers and tuckups, while the upper story of each is fitted up more or less comfortably for the use of the crews; gunning, fishing and camping outfits, with sails and gear, being kept there. On Sundays in particular the wharves and houses are crowded, the boats are off for short cruises up or down the river, or races are sailed between the recognized cracks, handled by old and skillful captains and trained crews. The following tables give very fully the details of the boats:

DIMENSIONS AND ELEMENTS OF TUCKUPS.

	Priscilla.	Igidious.
Length over all	15 02	15 04^4
l.w.l	14 11	....
Beam, extreme	4 03^6	4 05^6
l.w.l	3 07	3 08
Draft, bow	3	....
stern	10	....
Depth, amidship	1 05^2	1 04^3
Sheer, bow	6	7^6
stern	6	5^2
Displacement to l.w.l., lbs	716	....
to 7in. level line. lbs.	760	....
per inch immersion, lbs.	175	....
Area l.w. plane, sq. ft	43 52	....
lateral plane	8 00	....
centerboard	2 05	....
rudder	2 07	....
Total	13 02	....

AMATEUR CANOE BUILDING. 241

```
C.L.R. abaft fore end l.w.l. with board.  9 00      ...
C.E. abaft fore end l.w.l..............  8 27      ....
Station 0 to mast center..............  1 00      1 00
         slot in keel .........    { 4 00         50 8
                                    { 6 04         7 04
         point of coaming............  3 04       3 04
         fore thwart.................  6 02
         after thwart................  0 93
Mast, deck to truck...................15 00      15 00
    diameter at deck ................    3²       ....
                 truck...............    1⁶        3⁶
Boom.... ............................18 06      17 00
    diameter ........................    1⁷       ....
Gaff..................................10 00      10 00
    diameter ........................    1²       ....
Bowsprit, outboard....................11 11       2 06
Centerboard...........................            19×38
Mainsail, foot.......................18 00       16 00
         luff........................11 08       10 00
         head........................ 9 04        9 06
         leech.......................20 10       20 00
         tack to peak................19 10       19 00
         clew to thwart..............20 02       18 00
Area, sq. ft..........................  198        ...
```

Priscilla has a keel sided 5½in. at rabbet, 6½in. inside, 1in. thick, stem and sternpost sided 1in., transom ⅞in. thick, ribs 1×⅜in., spaced 9in., nails (copper riveted over burrs) spaced 3in. Planking, lapstrake, ⅜in., deck ½in., centerboard ¾in. oak, coaming ⅜in. oak, 3½in. high at point, ¾in. at midships and ⅝in. at stern. Round of deck, 1½in. Wearing strips, A A, oak, ⅝in.×⁷⁄₁₆in., spaced 6in. apart. Thwarts 7in. wide, 10in. above bottom of keel; trunk 11in. high.

Ingidious is 3ft. 2⅜in. wide across stern, with skag 3ft. 4in. long and 9½in. deep; coaming 4½in. high at point, 1in. from midships to stern. Keel 7¾in. extreme width; round of deck, 1¼in.; mast step of iron, braced with two rods with turnbuckles. Ribs and fastenings as in Priscilla. Planking ⅜in.

The boats are all lapstrake, very carefully built and copper-fastened, and are decked as shown, with about 7¼in. waterways, the well extending to the transom. The board is of the dagger pattern, often being much larger than shown, and the rudder is of the familiar barn door pattern, of great length, with tiller to match. The boards are always of wood, but at Norristown steel plates have lately been introduced, an innovation not approved of by the Philadelphia experts. Five metal boards of 50lbs. down, one a brass board, are now in use at Norristown. The sailing rules on the Delaware allow 4ft. 6in. beam for a 15ft. boat with five men all told, while the sail is limited to 56

TABLE OF OFFSETS—TUCKUP PRISCILLA.

Stations	Heights		Half-Breadths						
	Deck	Rabbet	Deck	12in.	8in.	6in.	4in.	2in.	Rabbet
0	1 9^2	....	0^4	0^4	......	......	......	......	......
1	1 7^5	1^4	7^6	5	3^2	2^6	1^6	0^6	0^4
2	1 6^2	0^3	1 1^2	9^7	7^1	5^6	4	2^1	0^7
3	1 5^2	....	1 5^3	1 2^3	11^2	9^2	6^6	3^7	1^4
4	1 4^4	....	1 8^4	1 6^1	1 3	1 1^7	9^6	6	1^7
5	1 3^6	....	1 10^5	1 9	1 6^2	1 4^2	1 1^7	8^1	2^2
6	1 3^3	....	2	1 10^7	1 8^4	1 6^6	1 3^3	10	2^3
7	1 3^2	....	2 0^7	2 0	1 9^7	1 8	1 4^6	11^2	2^4
8	1 3^3	....	2 1^3	2 5^4	1 10^4	1 8^4	1 5^2	11^6	2^4
9	1 3^5	....	2 1^3	2 0^4	1 10^2	1 8^3	1 4^7	10^7	2^4
10	1 4	....	2 0^7	1 11^7	1 9^2	1 7^2	1 3^3	9^1	2^3
11	1 4^5	....	2 0^2	1 10^5	1 7^2	1 4^6	1 0^4	6^6	2^1
12	1 5^5	0^2	1 11^3	1 8^4	1 4	1 1	8^6	4^2	1^6
13	1 6^6	1^4	1 9^7	1 5^4	11^6	8^2	4^5	1^5	1^2
14	1 8	4	1 8^1	1 1^7	6^3	3^1	0^7	......	0^7
15	1 9^3	8	1 5^6	5^4	0^4	......	......	......	0^4

linear feet of bolt rope when new, giving about 180ft. area. This will give 15ft. on foot, 13ft. luff, 8ft. head, and 21ft. leech. In the M. S. C. this rule is not used, the boats being classed together with a penalty for excess of sail area over that allowed. The limit is 165ft. for tuckups, 110ft. for duckers and 80ft. for canoes and small boats, the tuckups allowing the others five minutes over a five mile course. Any boat may increase her sail by allowing 2 seconds per foot per mile, and allowances are figured at the start, so that the first boat home wins. Five men are allowed to the tuckups and two for the duckers, but thus far a crew of three seems to be the best for the former.

Plate XLVII.—"Gracie," Open Sailing Boat.

Gracie is a rowboat 12ft. long and 34in. beam, designed and built by her owner, an amateur and a novice in boat building, Mr. E. A. Leopold of Norristown, Pa. The only guide, both in designing and building, was the first edition of "Canoe and Boat Building for Amateurs," and the boat was intended only for rowing and fishing on the Schuylkill River, a narrow and winding stream with very squally and variable winds, running through Norristown, a short distance from Philadelphia. The boat is a lapstreak, weighing 60lbs. when first completed, with a keel 1in. square, and fitted up with Allen's bow-facing oars. During the first year, 1885, she was used solely for pleasure rowing, duck shooting and fishing, and light enough to be handled conveniently. The next season a small sail was added, with several styles of leeboards, but the latter were in turn discarded for a variety of weather grip, while the sail grew to 59ft., some very fair sailing being done toward the latter part of the season. In 1887 the sail was increased to 85ft., as shown, while the weather grip was improved after many trials.

Thus rigged Gracie has raced against some fast boats of much greater size and power, and sailed by full professional crews, she sailing singlehanded and without ballast. She started thirteen times and won three firsts and three seconds in 1887, a very good record when the relative sizes and the reputation of the tuckups are considered. Of course she has had a good allowance, but it is hard to say whether it is too much all things considered. At first the crew of one sat on the floor, then two cushions were added, replaced a little later by a seat 5in. below gunwale, which seemed very high at first. It was soon evident that the crew could sit far out to windward, and with more comfort, while the pad, sewn in the back of the coat, to protect the backbone when sitting inside, was discarded. The next move was to place a seat, canoe fashion, across the gunwale, a very great improvement, as a better command of the boat is obtained with less effort. In sailing with but two sails the area was too small for racing, while the balance was bad, the boat constantly luffing. The addition of a jib cured both faults, the boat being faster except when free, and steering to perfection. In every puff she will eat herself to windward without a touch of the tiller, only a slight motion of the body forward or aft being necessary to luff her up or throw her head off. It is to this that she owes much of her gain, as little steering with the rudder is needed to make her work well in the constantly varying puffs that rush down from the hills in all directions. She is sailed without a fly, such as is used by most of the other boats, as it is very deceptive. In running free or in tacking the rudder is used, but most of the steering is done by the body only. The boat's worst point of sailing is before the wind, her best reaching. She receives about 9min. in five miles from boats 15ft. ×14ft. 6in., and sailed by four or five men hanging out to windward by means of ropes. The dimensions of Gracie are as follows:

Length extreme..	12ft.
Beam...	2ft. 10in.
Depth, amidships...................................	1ft.
at ends..	1ft. 8in.
Mainmast, from stem...............................	1ft.
above deck....................................	5ft. 10in.
diameter, deck...1½in., head........	1¼in.
Main boom..	8ft. 5in.
diameter......................................	1¾in.
Yard..	12ft. 4in.
diameter..........................1½in. and	¾in.
Batten...	8ft. 5in.
Mizenmast, above deck............................	8ft. 9in.

AMATEUR CANOE BUILDING. 245

Mizenmast, diameter..............1½in. and ¾in.
    from stem... .... 9ft. 7in.
Mizen boom................................... 5ft. 8in.
Batten... 4ft. 10in.
Bowsprit outboard........................... 4ft. 6½in.
Jib, or luff... .... 7ft. 6in.
    foot and leech, each..................... 5ft. 3in.

The jib halliard and downhaul are in one length, the bight belayed to a cleat on the port side, a small club is laced to the foot of the jib. The mainmast and boom are of bamboo, mainboom yellow pine, mizenmast of white pine. The mainsail is fitted to reef to a lateen by means of a jaw at B on the boom, so placed that no change of the halliard is necessary. The batten is fitted with cleats, C C C C, of spring brass, with a single reef point opposite each. The boom is lifted, a reef point made fast by one turn about the cleat, then the boom is shifted until the second jaw engages the mast. The other reef points may then be made fast at leisure, though in a short squall the jib is dropped, the mainboom made fast by but one reef point, and shifted to set by the inner jaw. The mizen is never reefed. In making the sails the spars were bent to position on the floor and the shape marked, then the stuff, a single width of sheeting, was cut and sewn. The weather grip, adopted after many experiments, is 3ft. 7in. on top, 2ft. on bottom and 1½in. deep, being immersed 10in. The top edge is ⅜in. thick, bottom ¼in. The distance from side is 2ft. 4in. and the immersed area 295sq. in. A keel has also been added, 4in. deep in all, of which the lower half is lead, 25lbs. The area of keel is 335sq. in., or with grip 630sq. in. The grip is hung from the sockets for the rowlocks by two cross pieces of wood in the form of an X, riveted where they cross and also to the top of the grip. The boat does not point as close as some of her competitors, but goes enough faster to make up for it, making sometimes five tacks to their four. The table of offsets is as follows, both ends being exactly alike:

TABLE OF OFFSETS—OPEN SAILING BOAT GRACIE.

Stations.	Deck, Height.	HALF-BREADTHS.				
		Deck.	No. 1.	No. 2.	No. 3.	No. 4.
0 and 12.........	1 8	0^4	0^1	0^1	....	....
1 and 11.........	1 5^3	5^4	3^2	2^4	1^6	0^7
2 and 10.........	1 3^3	9^6	7^2	6	4^5	2^5
3 and 9.........	1 2	1 0^6	11^2	10	8^2	5^2
4 and 8.........	1 1	1 3	1 2^1	1 1^2	11^6	8^6
5 and 7.........	1 0^2	1 4^3	1 4	1 3^2	1 2^2	11^7
6..................	1	1 5	1 4^6	1 4^1	1 3	1 0^6

In the winter 1887-8 a plate board of thin steel was added, the rig was changed to a single lug of 86ft., rigged as in Plate XLVII., and a light horizontal wheel was fitted directly on the rudder head, in place of the tiller, all these changes being for the better.

PLATE XLVIII.—"CLIO," HALF DECKED SAILING BOAT.

Toronto Bay, on Lake Ontario, is the home of a fleet of small boats, and much racing is done there through the season. Clio was the champion in 1887 and is a good example of her class. She is of pine, lapstreak, of $\frac{5}{16}$in. plank, and is 16ft. long, 3ft. 8in. beam, and decked for 4ft. 6in. forward, 2ft. aft, and with waterways of 4 to 6in., the coaming being 3in. high. The leading dimensions are:

  Mast, deck to head........................ 16ft.
   from stem............................ 3ft.
   diameter, deck 5in................head 1in.
  Bowsprit, outboard......................... 5ft. 6in.
  Boom...................................... 18ft.
  Yard...................................... 10ft. 6in.
  Spinaker boom............................ 15ft.
   hoist................................ 14ft.

AMATEUR CANOE BUILDING. 247

```
Jib, luff......................................... 14ft.
     foot......................................... 8ft.
     leech........................................ 12ft.
     area, square feet............................ 40
Mainsail, foot................................... 17ft. 6in.
         luff.................................... 10ft.
         head.................................... 10ft.
         leech................................... 10ft. 9in.
         tack to peak............................ 19ft. 6in.
         clew to throat.......................... 19ft. 6in.
         area, square feet....................... 190
```

The sails are of light drill, the roping on leech only extending as high as the reefs. The centerboard is of $\frac{3}{16}$in. iron, weighing 85lbs., 4ft. long and 3ft. 3in. deep. It is sharpened on the forward edge. The boat is double-ended, so the main sheet works on a high traveler over the tiller.

PLATE XLIX.—SMALL DINGEYS.

The larger boat shown in Plate XLIX. is a cross between a canoe and a sneakbox, intended as a tender to a small yacht, the object being to obtain something narrow enough to fit into the yacht's gangway, ready for use at all times and also stiff and safe. The features of this boat are a "shovel nose" to facilitate towing when preferred, rather small beam, well held fore and aft, long flat floor, quick bilge and high side with a light deck and coaming in canoe fashion. The shape of the moulds at three cross sections are shown by the dotted lines. She is 9ft. long, 2¼ft. wide, 1ft. deep in center with a sheer of 2in., and supplied with an iron centerboard and triangular sail 7ft. on foot, head and leech. The board is of ¼in. boiler iron with 1ft. vertical drop. The sail is set upon a short stump pole after the plan of the Lord Ross lateen for canoes.

A boat of the ordinary style, but extremely serviceable, is also shown in the diagrams. From these it will be seen she possesses great width, with long, flat floor and high sides, tumbling home at the stern and along the side. This tender is remarkable for the load she carries and for her stiffness, which makes her a more reliable and useful adjunct than many dingeys twice the length. She is only 6ft. 6in. long over all, with an extreme beam of 3ft. 1½in. Her fault is towing heavily when sailing fast, and the difficulty of stowing on deck on account of her width.

PLATE L.—SPORTING BOATS.

The larger drawing represents a boat for ducking and shooting, thus described by her builder:

She shows but little above the water, draws but little, and so can be used in shoal water, can easily be transformed into a capital blind by using a little grass, weed, or brush on the deck. She is not easily turned over, and a person can shoot from any position in her, which he cannot do in a canoe. I know this from experience, as I have spent many a day in one.

In the first place, to get frames or ribs lay out on the floor a cross section both ways of the boat, full size; lay off the ribs or frames a foot apart the whole length, and taking the measure of each one on the horizontal plan gives you the length, and in the perpendicular section the breadth. Then on the ends leave the width of the sides, which in my boat is only 12½in. Then take a strip of thin stuff, and from a dot that you make for the width on each side of the center spring the strip to the width of sides at each end, top and bottom, and you have the curves for the ribs. Saw out the center as far as the cockpit comes, and you have the forms. Stay them to the floor, and put on the bottom first.

Material for frames and ribs ½in. oak, also for the sides, which are only 1½in. wide. Screw the sides to the ribs, stem and sternpost with ⅞in. No. 6 wire brass screws. It is now ready for the bottom. Use $\tfrac{3}{16}$in. oak ripped to 6in. in width, and where each joint comes use a batten $\tfrac{3}{16}$ by 1½in., clinched through about 1½in. apart with brass escutcheon pins, driving them through on the face of a hammer or piece of iron.

Use plenty of white lead on the battens and on the edge of the sides. Fasten the covering to the ribs and sides with $\frac{5}{8}$in. No. 1 wire screws and escutcheon pins.

For floor to the cockpit use $\frac{5}{8}$in. pine, and the washboards to cockpit $\frac{5}{8}$in. black walnut worked up and down and screwed to a strip let into the top of the frames, and at the bottom by strips put between the ribs. The midship section (No. 3) shows it in detail.

Amount of material: about 100ft. of $\frac{3}{16}$in. oak, 20ft. of $\frac{1}{4}$in. oak, enough $\frac{7}{8}$in. oak for stem, sternpost and keel, 12ft. of $\frac{5}{8}$in. pine for floor, and enough material for the washboards, which can be black walnut, pine, oak, or whatever a person chooses, 1$\frac{1}{2}$ gross of $\frac{7}{8}$in. No. 6 wire brass scews, 3 gross of $\frac{5}{8}$in. No. 1 wire brass screws, and 8oz. of stout brass escutcheon pins $\frac{1}{2}$in long. This is all that is required but paint.

I shall rig my boat to sail, using two legs of mutton sails, sharpie rig, and also to row. The oarlocks will want to be placed on the outside and raised up high enough to clear the washboards, which can be done by a block or an iron, the boat being so wide it can be used with quite a long oar, and by a good oarsman it can be sent along very fast.

The smaller boat is for a similar purpose, but is built of canvas, as follows: The ribs and long strips are made of oak $\frac{3}{8}$ by scant $\frac{1}{4}$in.; the ribs are placed 5in. apart, and there are six long strips on each side, and two more 8ft. long to fill up the larger space in the middle. Where each strip and rib cross they are clinched together with a copper nail. The gunwale strips are $\frac{3}{8}$ square, and each rib is let into them and nailed with two copper nails. Bring the canvas over the dado in the stem and stern, and put in a spline; then put on a keel made of oak outside of the canvas and screw it to the center keelson. The cockpit is made of half-inch black walnut screwed to the gunwale strips, and has a piece $\frac{1}{4}$ by 1in. screwed on top on the sides and back, so that it leaves $\frac{1}{2}$in. projection. In front use a piece $\frac{1}{2}$in. by 3in.

The seat is made of two $\frac{3}{8}$in. pine pieces, 3in. wide, screwed to the ribs, and the top is rabbeted $\frac{1}{2}$x$\frac{1}{2}$, and the top is made of

2in. by ¼in. pine strips placed 1in. apart and cleated together.

The deck is raised 1in. in center of boat, so that it sheds the water, both sideways and endways.

Bring the canvas around the boat and nail it on top of gunwale, and the deck the same, and then put a neat ¼in. half-round moulding on top of the tacks, so that it makes a neat job.

FLAT-BOTTOMED BOATS.

TO BUILD a framed boat with a round bottom requires time, skilled labor and good material, but there are many cases where a boat is desired for temporary use, for hard work where a light boat would soon be destroyed, or in a hurry, in which cases beauty, light weight and speed are of little importance, the requisites being carrying capacity, cheapness, and a saving of time. In such cases the methods previously described are not applicable, but the ends in view will be best filled by some variety of "flat-bottomed" boat, as they are commonly called. With the rougher of these craft but little skill is required to turn out a strong and useful boat, the operations being little more difficult than the making of a common box, and even with the finer boats of this class no special skill is needed beyond the ability to use the ordinary tools of the house carpenter. While flat-bottomed boats are usually heavy clumsy and ill-shaped, there is no reason why they may not, with care and a little skill, be almost as light and shapely and for many purposes as good or even better than the more costly lapstreak or carvel built craft.

The cheapest and simplest of all boats is the scow (Fig. 1), a style of boat that may be built in a few hours and at an expense of two or three dollars only. In almost all places a few common boards of pine, spruce, or almost any wood, can readily be obtained, the commonest size in America being 13ft. long, 10in. wide and 1in. thick. To construct a boat from such material to carry two or three persons, four or five boards will be necessary. Two of these should be selected and a length of 10ft. sawn from each. The edges of these pieces are now planed or "jointed" up straight and square to the sides, the latter being either planed or left rough. These two side pieces (*a*) are laid one on the other, and two or three small nails driven through them to hold them temporarily together, and the outline of the side is now marked on the upper one. The upper edge of the boat will be straight, the bottom will be straight for 5ft. amidships, and at each end for 2ft. 6in. will slant upward until the end pieces of the boat (*b b*) are but 4in. deep. The two boards are now sawn to shape and planed square on the ends and the slanting portions of the bottom, then they may be taken apart.

Each end piece will be 3ft. long, or longer if a wider boat is required, and 4½in. wide in the rough. The upper edges

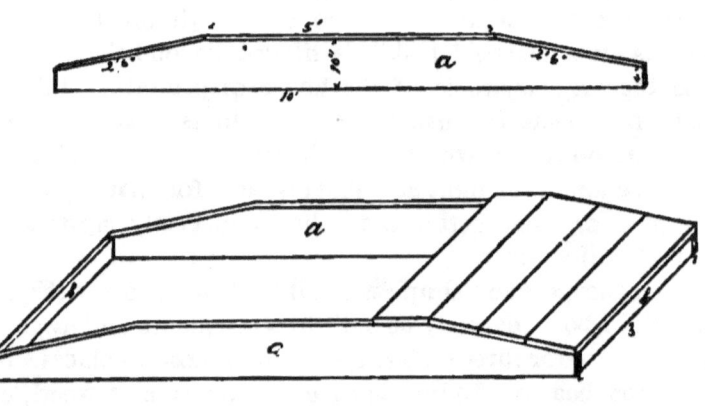

FIG. 1—SCOW.

are planed up, and the sides are each nailed to the ends, using eight-penny nails, or ten-penny if the stuff is over ¾in. thick. The frame is now turned bottom up, the two end pieces are planed on their bottom edges to correspond with the bevel of the bottom, then a sufficient number of pieces to cover the bottom are sawn off the remaining boards. In this case they will each be 3ft. 2in. long. Their edges are carefully "jointed up" straight and square, and they are nailed in place across the bottom. When all are nailed on the ends may be planed down even with the sides of the boat. To stiffen the bottom a strip 5in. wide and ¾ to 1in. thick (see *i*, Fig. 2) is laid down the center of the bottom inside and nailed with wrought or clinch nails to each plank, the nails being driven through and their points clinched or turned in, using a hammer and an iron set. About 2ft. at each end will be covered with a deck, as at *h*, Fig. 2. One seat will be put in for rowing, being supported on two cleats, one nailed to each side. Iron rowlocks may be obtained in most localities at a cost of seventy-five cents per pair, and are better than wooden ones, but if they are not to be had, the latter can be made of oak. A cleat of oak 1¼in. thick, 2in. deep and 9in. long is screwed along the inside of the gunwale. In each cleat two mortises are cut, 1½in. long, ½in. wide, and 3¼in. apart. The rowlocks are each 7in. long, ½in. thick, 2in. wide above the cleat, and 1½in. wide in the mortises, projecting 4in. above the gunwale and 3in. below.

If all the joints are neatly made, the boat should be tight after being in the water a short time; but it is always best to paint or tar the entire boat, inside and out, preserving the wood and lessening the chance of leakage. In no case should caulking be needed in a new boat. If the builder desires, each edge can be painted as the board is put in place, which will still further prevent any leakage.

While such a boat is often all that is needed, with a little more care and skill a much better one may be built. The punt, as it is commonly called (Fig. 2), is a scow of rather better design than the one described above, but the operations of building are similar. These boats are often used for fishing on rivers and ponds, as they are roomy, stiff and safe from any danger of capsizing, and the occupants can sit all day in comfort, or move about freely, which cannot be done in a round-bottomed boat of similar size. Such a boat may be 14 to 16ft. long, 4ft. beam at gunwale, 3ft. 4in. at bottom, and the sides 14in. deep. The sides (*ff*) will each be a little longer than the length of the completed boat, 14in. wide and ¾in. thick. They should be free from knots and sap wood, and as nearly alike as possible, so as to bend equally. One is laid on two benches, the outline of the boat is marked out as shown, the ends sweeping upward in easy curves, and it is sawn and planed to shape. It is then laid on the second board, the two are lightly nailed together, and the latter planed to match, a center line being marked on both while nailed together. The two end pieces (*c c*) are next sawn out of 1in. oak or ash, the ends being beveled, as the bottom of the boat througout will be narrower than the top. Next a piece (*d*) 14 to 16in. wide and 4ft. long is sawn off and the ends beveled, making it 4ft. long on the upper edge and 3ft. 4in. near the lower. The two small projections (*e e*) are left, to aid in setting the

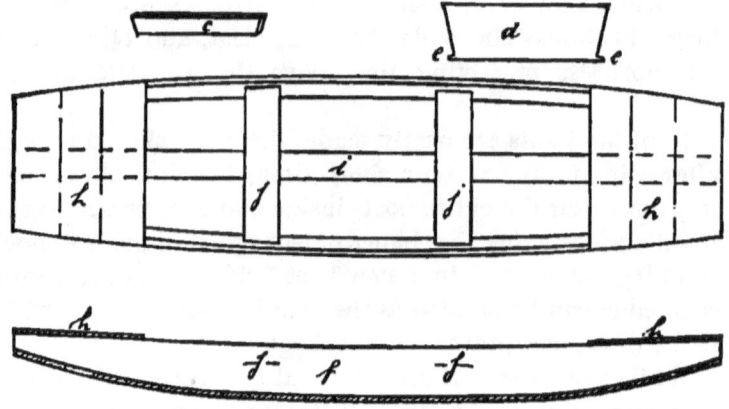

Fig. 2—Punt.

side correctly. This board or mould is placed on edge, one side board is laid in place against it at the center mark, and a few nails are driven through the side board into the end of the piece. Now the other side is fitted in the same manner. The three pieces resting on a level floor the corresponding ends of the side pieces are drawn together with ropes until the end pieces will just fit between, then the sides are nailed or screwed to the ends. The best way to do this is to bore the holes and fit each side in turn to its corresponding end piece, putting in the screws, before the sides are nailed to the mould (the pieces after fitting being taken apart); then when the ends are finally in place there is no trouble in holding and adjusting them, the screws being reinserted in the holes already bored.

When sides and ends are well fastened together, both of the frames should have the same degree of curve, and the entire frame should be true and symmetrical. The lower edges of the sides having been planed square, now require to be beveled slightly, on account of the outward flare of the sides. To do this a piece of board, one of those cut for the bottom, is laid across and used as a guide, the outer corner of each edge, both of sides and ends, being planed off until the board lies flat across all the edges. The bottom boards are now cut to length and nailed in place, the edges of each being very carefully planed up to fit its neighbors. When the bottom is on, the ends are planed off even with the side of the boat, it is turned over and a strip (i) 5in. wide is nailed down the middle of the bottom, as in the previous boat. This strip will be 1in. thick at its center, but toward the ends it may diminish to $\frac{1}{4}$in. so as to bend more easily to the curve of the bottom. When it is in, the ends are decked over for two or three feet, as at $h\,h$. Two thwarts or seats ($j\,j$) will be put in, each 9in. wide and 1in. thick. They should be placed about 7in. below the gunwale, and each end will rest on a short piece nailed to the side of the boat, long enough to reach from the bottom to the wider side of the seat. The seats should be secured well to the sides, as they serve to stiffen the boat. A gunwale strip is usually run around the outer edge. It may be of oak $\frac{4}{4}$in. wide and 1$\frac{1}{4}$in. thick, screwed to the side pieces. Rowlocks and stretchers complete the boat. It will, however, be easier to row straight if a skag be added to the after end. A stern post of oak 1x1$\frac{1}{4}$in. is nailed down the center of the end, and in the angle between it and the bottom is fitted a piece of 1in. board (o, Fig. 3) nailed to it and the bottom. On the stern post a rudder may be hung if desired.

While such a boat answers very well for fishing and similar purposes, if much rowing or sailing is done, a better form

256　　　*FLAT-BOTTOMED BOATS.*

is that of the skiff or bateau shown in Fig. 3. In this boat the after end is similar to the previous one, but the bow is very different, resembling more a round-bottomed boat. The

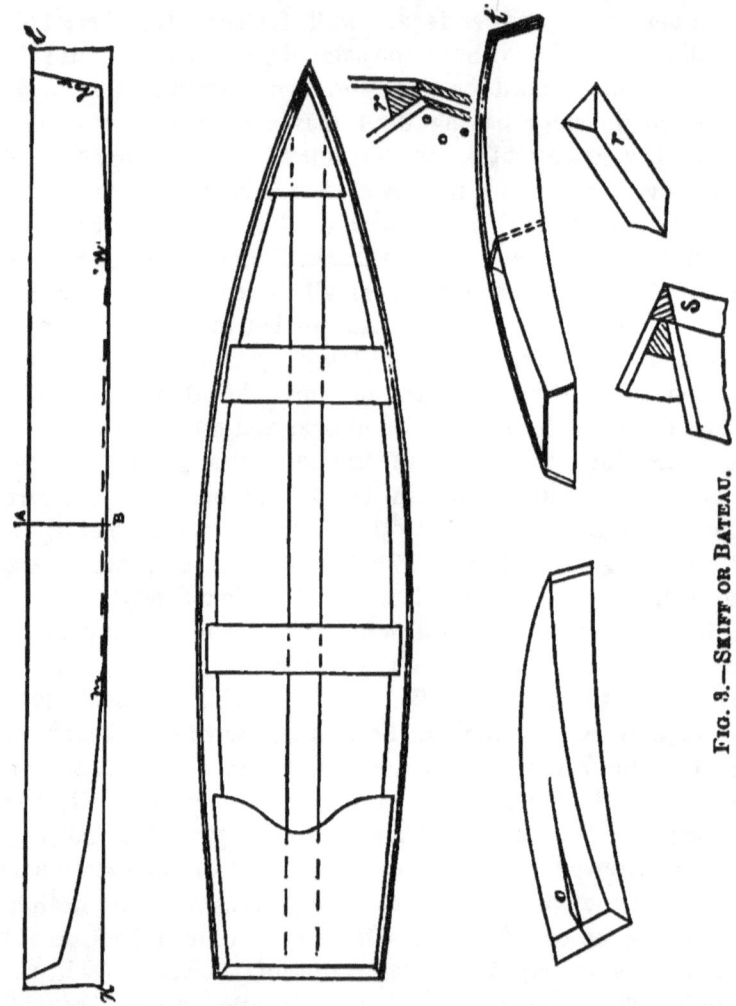

Fig. 3.—Skiff or Bateau.

sideboards are marked and cut as in the former boat, but at the fore-end they are not cut up at all, but are sawn off at a slight bevel to fit the forward rake of the stem (*k l* shows the sideboard in the rough, with the side marked out). The gunwale will have a slight sheer, part of it being due to the bending of the sideboards, but to increase it the upper edges are made a little hollow, their concavity being from 1 to 2in., according to the sheer desired. A middle mould is cut out similar to d, and also a stern piece, the latter of 1in. oak. It is fitted and screwed to each sideboard in turn, then it is taken off, the sideboards are nailed to the mould along the lines A B, and the sternboard is replaced and screwed fast. Now the two sides are drawn together with a rope at their fore-ends until they nearly or quite meet, as at t, and a piece of oak of triangular form (r) is cut to fit in the angle between them, and they are screwed fast to it. The bend of the sides will cause the bottom of the boat to have considerable rocker, usually much more than is desirable. To avoid this, when the frame is thus far completed, the bottom edges of both sideboards are planed down from m to n, until the bottom is straight for some distance amidships. This can best be determined by setting the frame, top upward, on a level floor. When the edges are planed off equally they must be beveled, as in the preceding boat, the floor is nailed on, the middle piece is put in and nailed down, and the thwarts put in. Both in bow and stern there will also be seats at about 3in. below the gunwale and of the shape shown. To complete the bow, the ends of the sideboards are planed off, and another triangular piece of oak (s) is sawn out and nailed against the ends and the piece r, as shown, making a sharp bow. A scag (o) is also added, wale strips are put on, and the boat is ready for painting. Such a boat may have a centerboard, as described in the previous chapters, and may also be fitted with sails in the same manner as an ordinary round-bottomed boat.

THE DORY.

These boats are largely used by the fishermen of the Atlantic Coast, both along shore and on the fishing vessels, and they are also suitable for rowing and as service boats for yachts.

The boat here shown is planked with white pine, the sides, of three pieces each, being $\frac{9}{16}$in., and the bottom $\frac{7}{8}$in. The laps of the siding are rabbeted, so as to make a flush surface inside and out. The timbers are of oak $1\frac{3}{8} \times$ 1in. and $2\frac{1}{2}$in. in thwarts. The gunwales are $1\frac{1}{2}$in. $\times 1\frac{1}{4}$in., with a $\frac{5}{8}$in. strip on top, covering edge of upper streak also. There are three movable thwarts resting on risings, and removed when the boats are nested or packed. The sizes are so arranged that five boats can be stored together, one within the other, thus occupying little space on deck.

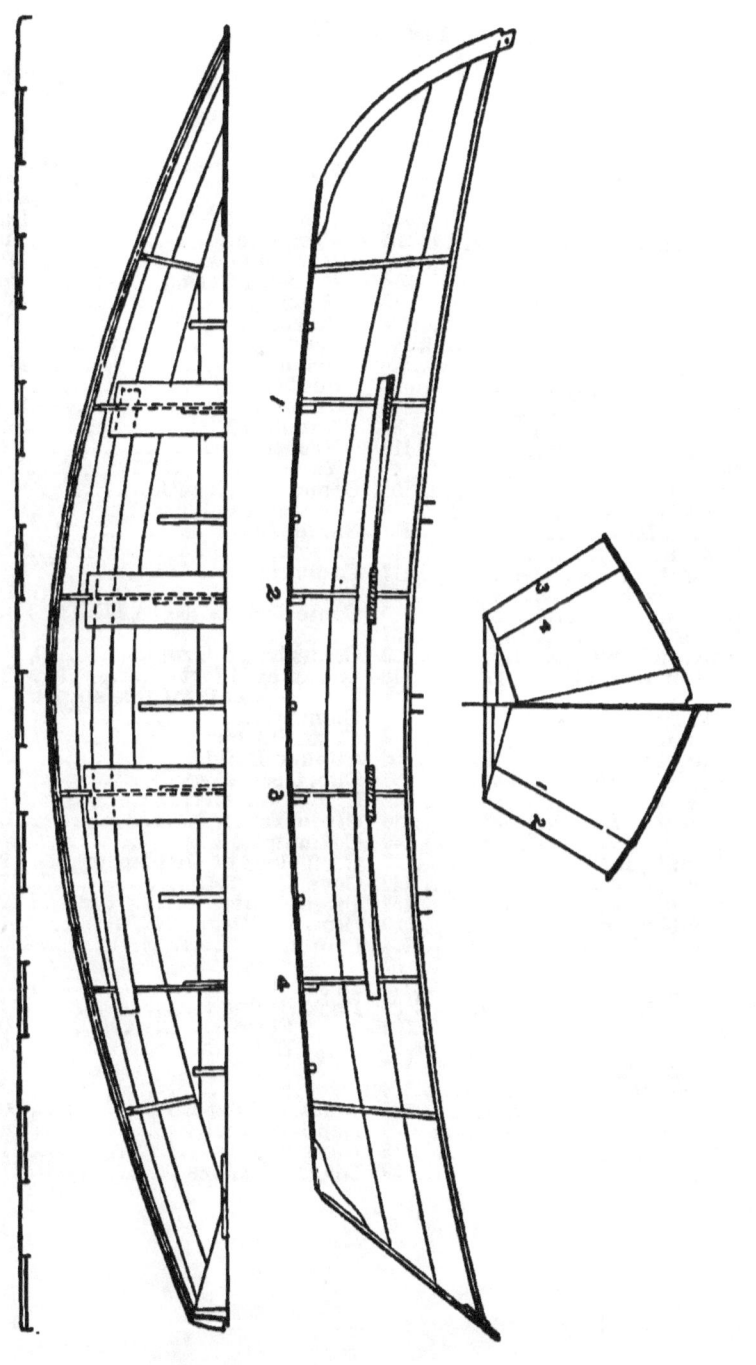

INDEX.

	Page
Annie, Canoe Yawl	205
Apron	62, 115
Back Rabbet	42
Backboard Canoe	66
Boat	119
Balance Lug	84, 166, 225
Ballast in Canoes	161
Barnegat Cruiser	216
Sneakbox	215
Bateau	251
Battens	84, 141
Beam	126
Beading Line	42
Beds, Camp	104
Bench	34
Bending Timbers	119
Bevel	49
Bilge Keels	55
Block Model	19
Boat, Flat	167
Body Plan	12
Breasthook	117
Broadstreak	49
Building	27, 40
Burrs	52
Buttock Line	12, 18, 123
Button Boards	118

CANOE:
Canadian	7
Canvas	80, 111, 156
Classification	8, 125
Clyde	141
Cruiser, Barnegat	216
Cruising	137, 141, 145, 149
Definition	7
Dot	139
Double	153
Doubleskin	29
Elements	124
Fittings	57
Guenn	179
Herald	29
Ione	202
Jacket	62, 64
Jersey Blue	13, 22, 187
Lacowsic	174
Laloo	141
Lassie	165
Mersey	212
Metal	29

	Page
CANOE:	
Nautilus	145
Notus	176
Ontario	30
Paper	30
Pearl	147, 158
Pecowsic	170
Raritania	137
Ribband Carvel	28
River	137
Sails	70
Seat	66
Shadow	139
Sunbeam	169
Tandem	153
Vagabond	200
Vesper	174
Yawl	204
Canoe Construction	180
Canvas Boat	165
Carvel Build	27, 119, 233
Cassy	206
Caulking	27, 121
Cedar	132
Centerboards	88, 93, 125, 127, 134, 187
Centerboard Trunks	40, 90
Center of Effort	71
Lateral Resistance	71
Clamps	34
Cleat, Butler	190
Clinker Build	27
Clio, Open Boat	246
Clyde Canoe	141
Clyde Tent	103
Coamings	55
Coefficient of Displacement	9
Cove	117
Cross Spalls	52
Crown of Deck	54
Crutch	129
Decks	54
Deck Beams	54, 112
Flaps	60
Hatches	66
Tiller	131, 190
Yoke	129
Delta, Sailing Boat	230
Depth of Keel	126
Designing	11
Diagonal Lines	12, 17

INDEX. 261

	Page
Dingeys	247
Displacement	9
Dory	258
Draft	10
Drawing Instruments	11
Paper	12
Drip Cups	68
Drop Rudder	95, 190
Ducker	237
Fairing	17
Farnham's Apron	63, 133
Finishing	56
Fittings	57
Flaps, Deck	60
Floor Boards	53, 54, 118
Knees	117
Footlines	118
Foot Yoke	129
Forest and Stream Cruiser	221
Freeboard	10
Garboard	45
Gracie	243
Gridiron	107
Gunwale	112, 117
Half Breadth Plan	12
Halliard	82, 141
Hatches	66, 149
Headledges	43, 90
Hunting Boats	248
Inwale	112
Iris	213
Jackstay	86
Keel	40, 112, 115, 126
Batten	40
Keelson	115
Kittiwake Tent	102
Lamps	110
Lapstreak	27
Laying Down	21
Ledges	53
Limber Holes	53
Lines	12
Mast Steps and Tubes	55
Materials	31
Mattress	104
Measurement Rules	10, 125
Mess Chest	110
Middle Ordinate	16, 123
Midship Section	10, 14
Models	8, 19
Mohican Sail	159
Moulds	24, 115

	Page
Nailing Plank	50, 51
Offsets, Tables of	21
Oliver Lateen	78
Paddles	67
Painting	56
Panels	141
Paper Canoes	30
Pecowsic	170
Planking	32, 45
Plank, Taking Off	23
Punts	251
Rabbet Line	40, 115
Racks	118
Reefing Gear	78, 141, 145, 198
Ribs	44, 52, 112
Rib and Batten Canoes	28
Risings	118
Riveting	52
Rocker	10
Rowboats	115, 122, 156
Rowlocks	118
Rudders	94, 119, 189, 199
Fastenings	90
Lines	132
Sails	70
Canoe Yawl	208
Clio	246
Delta	256
Dot	140
Forest & Stream Cruiser	225
Gracie	244
Ione	203
Iris	214
Laloo	143
Lassie	177
Lateen	70
Leg of Mutton	76
Lord Ross Lateen	76
Lugs	82, 166, 169
Measurement of	72
Mohican	78, 159
Notus	178
Oliver Lateen	78
Pecowsic	173
Sea Bee	180
Sharpie	76
Sneakbox	166
Stevens	182
Sunbeam	163
Tandem Canoe	155
Vagabond	192
Vesper	176
Vital Spark	203
Yawl	160
Scag Band	119
Scow	251
Seats, Canoe	66

	Page		Page
Set	34	Telescopic Apron	63
Sharpie Rig	76	Tents	100, 133
Sheer	10	Thwarts	118
Shutter	121	Tiller	131, 190
Sirmark	46, 48	Timbers, Bending	119
Skirt Jacket	62, 64	Timber Block	119
Sneakbox	215	Timbering Canoe	52
Spiling	46, 119	Tools	33
Spinaker	193	Transom Knees	117
Staff	46	Trunks	40
Stem	40, 115	Tuckup	239
Band	55, 119		
Steam Box	59, 119	Upper Streak	45, 117
Steering Gears	123, 159		
Stephens' Centerboard	187	Varnishing	56
Rudder Gear	189	Vagabond	200
Stern	116	Vesper	174
Stirrups	120, 132	Vise	34
Stocks	36	Vital Spark	207
Stopwaters	44		
Stretcher	130	Watertight Hatches	149
Stoves	106, 108	Weather Helm	74
		Wells	55, 57
Tabernacles	97, 209	Work Bench	34
Tack	82, 141		
Tiller, Deck	131, 190	Yachts' Boats	163
Butler	199	Yawl, Canoe	160, 187
Topping Lift	86	Yoke	67

LIST OF PLATES.

    I. Cruising Canoe "Jersey Blue"—Lines.
   II. Canoe "Jersey Blue"—Construction Drawing.
  III. River Canee "Raritania."
  IV. Cruising Canoe "Dot"—Shadow Model.
   V. "Dot"—Racing Sail.
  VI. Clyde Canoe "Laloo."
 VII. "Laloo"—Sail Plan.
VIII. Racing and Cruising Canoe "Nautilus."
  IX. Cruising Canoe "Pearl" No. 3.
   X. Racing Canoe "Pearl" No. 6.
  XI. American Cruising and Racing Canoe.
 XII. Tandem Canoe.
XIII. Tandem Canoe—Sail Plan.
XIV. Canvas Canoe.
 XV. Fourteen-foot Rowboat.
XVI. Rowing and Sailing Boat.
XVII. Mohican Sail. Canoe Footgear.

LIST OF PLATES.

- XVIII. Canoe Fittings.
- XIX. Class A Canoe "Lassie."
- XX. "Lassie"—Sail Plan.
- XXI. Class B Canoe "Sunbeam."
- XXII. Class B Canoe "Pecowsic."
- XXIII. "Pecowsic"—Sail Plan. "Vesper"—Sail Plan.
- XXIV. Class B Canoe "Vesper."
- XXV. Class B Canoe "Notus."
- XXVI. "Notus"—Sail Plan.
- XXVII. Class B Canoe "Guenn"—Lines.
- XXVIII. Class B Canoe "Guenn"—Fittings.
- XXIX. Class B Canoe "Guenn"—Sail Plan. Drop Rudder.
- XXIXa. Canoe Fittings.
- XXX. Class B Canoe "Vagabond."
- XXXa. Body Plans of "Vesper" and "Vagabond." "Iris" Sail Plan.
- XXXI. Canoe "Ione."
- XXXII. "Ione"—Sail Plan. Canoe Yawl "Annie."
- XXXIII. Canoe Yawl "Cassy."
- XXXIV. Canoe Yawl "Vital Spark."
- XXXV. "Vital Spark"—Sail Plan.
- XXXVI. Mersey Canoe.
- XXXVII. Canoe Yawl "Iris." (Sail Plan on Plate XXXa.)
- XXXVIII. Barnegat Sneakbox.
- XXXIX. Barnegat Cruiser.
- XL. "Forest and Stream" Cruiser.
- XLI. "Forest and Stream" Cruiser—Sail Plan.
- XLII. Sailing and Cruising Boat "Delta."
- XLIII. Sailing and Cruising Boat "Delta"—Sail Plan.
- XLIV. Delaware River "Ducker."
- XLV. Delaware River "Tuckup."
- XLVI. "Tuckup" Sail Plan. "Forest and Stream" Cruiser—Body Plan.
- XLVII. Sailing Skiff "Gracie."
- XLVIII. Sailing Skiff "Clio."
- XLIX. Small Dingeys.
- L. Sporting Boats.

R. J. Douglas & Co.,
WAUKEGAN, ILL.

CANOES.

SMALL CRAFT OF ALL TYPES.

Furniture and Fittings of all kinds.

WORKMANSHIP AND MODELS UNSURPASSED.

Prices Moderate.

Send for our illustrated catalogue of books on fishing, shooting, canoeing, yachting, camping, dogs, natural history, outdoor life, field sports, travel, adventure, etc., etc. Free to any address. FOREST AND STREAM PUBLISHING COMPANY, 318 Broadway, New York.

St. Marys, O.,
Montpelier, O. } OAR MILLS. NEPTUNE ANCHOR WORKS.

De Grauw, Aymar & Co.,

Manufacturers and Importers of

Cordage, Oakum, Wire Rope, Chains, Anchors, Oars, Blocks,

Buntings, Flags, Cotton & Flax Ducks,

Russia Bolt Rope, Marine Hardware,

AND

Ship Chandlers' Goods Generally.

34 & 35 SOUTH STREET, NEW YORK.

Orange Sporting Powder.

(ORANGE MILLS, Established 1808.)

MANUFACTURED BY

Laflin & Rand Powder Co.

ORANGE RIFLE,
   ORANGE SPECIAL,
       ORANGE DUCKING,
           ORANGE LIGHTNING.

The Most Popular Powder in Use. Of Superior Excellence.

SEND POSTAL CARD FOR ILLUSTRATED PAMPHLET SHOWING SIZES OF GRAINS. MAILED FREE.

NEW YORK OFFICE: 29 MURRAY STREET.

BRANCH OFFICES:

St. Louis, Mo.; Chicago, Ill.; Dubuque, Iowa.; Cincinnati, Ohio; Baltimore, Md.; Pittsburgh, Pa.; Denver, Col.

For sale generally throughout the United States.

www.ingramcontent.com/pod-product-compliance
Lightning Source LLC
Chambersburg PA
CBHW032133230426
43672CB00011B/2322